Gregory Matoesian, Kristin Enola Gilbert
Practicing Linguistics Without a License

Foundations in Language and Law

Editors
Janet Giltrow
Dieter Stein

Volume 9

Gregory Matoesian, Kristin Enola Gilbert

Practicing Linguistics Without a License

Multimodal Oratory in Legal Performance

DE GRUYTER
MOUTON

ISBN 978-3-11-162823-3
e-ISBN (PDF) 978-3-11-102557-5
e-ISBN (EPUB) 978-3-11-102582-7
ISSN 2627-3950

Library of Congress Control Number: 2023934183

Bibliographic information published by the Deutsche Nationalbibliothek
The Deutsche Nationalbibliothek lists this publication in the Deutsche Nationalbibliografie;
detailed bibliographic data are available on the internet at http://dnb.dnb.de.

© 2024 Walter de Gruyter GmbH, Berlin/Boston
This volume is text- and page-identical with the hardback published in 2023.
Cover image: kokouu/E+/Getty Images
Typesetting: Integra Software Services Pvt. Ltd.

www.degruyter.com

Acknowledgments

Thanks to numerous colleagues who have provided help over the years: Vainis Aleksa, Krisda Chaemsaithong, Alan Durant, Brigittine French, Simon Harrison, Adam Kendon, Michael Lempert, Janny Leung, Rubert Stasch, Dieter Stein, Lynn Taylor and Teun van Dijk for their input on technical matters. We would also like to thank our friends for support: Maurice Rottmann, Stephanie Dressler, Ronjon Paul and Adam Mollsen. Thanks to Sage, Routledge, Science Direct, and Cambridge University Press for permission to use parts of prior works. Finally and above all, we thank Michael Silverstein for his support over the years, and we are deeply indebted to his seminal scholarship.

Contents

Acknowledgments —— V

List of figures —— IX

1	**Introduction —— 1**	
1.1	A distinct speech-exchange system? —— 2	
1.2	Multimodal conduct —— 6	
1.2.1	Gesture —— 6	
1.2.2	Facial expressions, gaze, movement, and material objects —— 10	
1.3	Poetics and affect in legal oratory —— 11	
1.4	Data and methodology —— 13	
1.5	Chapter overview —— 15	
2	***This is not a course in trial practice:* Multimodal participation in objections —— 18**	
2.1	Literature review —— 19	
2.2	Embodied participation and affective stance —— 23	
2.3	Objection conference: Cross-examination within cross-examination —— 32	
2.4	Summary —— 40	
3	***She does not flee the house:* Poetic rhythms of space, path, and motion —— 42**	
3.1	Multimodal conduct in language and law —— 43	
3.2	Introducing and contextualizing the exhibit —— 44	
3.3	A multimodal poetics of space, path, and motion in three parts —— 49	
3.3.1	Staying in the house: Departure from normative expectations (lines 001–005) —— 49	
3.3.2	The back door: A crucial parenthetical (lines 006–011) —— 55	
3.3.3	Rummages around the property: Tracing motion events (lines 012–016) —— 59	
3.4	Summary —— 62	
4	**Historical voices, collective memory, and interdiscursive trauma in the legal order —— 64**	
4.1	What is collective memory? —— 65	
4.2	Gesture in microcosmic action —— 66	

4.3	A "strange" feature of the legal case —— 67	
4.4	Poetic oratory as microcosmic ritual —— 70	
4.5	Section summary —— 79	
4.6	The interactive co-construction of collective memory and cultural trauma —— 81	
4.7	Summary —— 88	
5	**Language, gesture and power in closing argument** —— 90	
5.1	The relevance of gesture and multimodal conduct for closing —— 91	
5.1.1	Beat gestures —— 91	
5.2	Background themes —— 92	
5.3	Interdigital beats and obstructing justice —— 92	
5.4	Interdigital beats and the embodied resistance ideology —— 97	
5.5	Type-token reflexivity —— 104	
5.6	Multimodal modulations —— 107	
5.7	Residual semanticity —— 108	
5.8	Power and multimodal conduct in the law —— 109	
6	*Unless he has three or four arms:* **Enacting evidence** —— 112	
6.1	Demonstrations of evidence —— 113	
6.2	Anytime we talk about sex it's embarrassing —— 114	
6.3	He does not have enough arms and hands to do that —— 119	
6.4	Summary —— 129	
7	**Conclusion** —— 132	
7.1	Rape reform —— 135	
7.1.1	Pre-recorded questions and neutral translator —— 136	
7.1.2	Seeking truth and the accuracy of evidence —— 139	
7.1.3	Rape myths —— 141	

Transcription conventions used —— 143

References —— 145

Index —— 153

List of figures

Figure 1.1   *who* interdigital upstroke —— 9
Figure 1.2   *is* interdigital downstroke —— 9
Figure 1.3   *a hundred and ninety-five pounds* intradigital x4 —— 9
Figure 1.4   *as Miss Lasch* parenthetical or PP —— 10
Figure 1.5   *eleven and a half shoe* interdigital downstroke —— 10
Figure 2.1   line 46 *open mouth suspension* —— 26
Figure 2.2   line 46 *body recoil, head tilt, pursed lips and closed eyes* —— 26
Figure 2.3   line 60 *J closed mouth gaze at PA* —— 28
Figure 2.4   line 61 *open mouth gaze at DA* —— 29
Figure 2.5   lines 25–27 *precision-ring gesture* —— 34
Figure 2.6   line 33 *CR scrolling up steno paper* —— 35
Figure 2.7   line 62 *closed eyes in lateral head shake* —— 37
Figure 2.8   line 68 *head tilt* —— 38
Figure 2.9   line 71 *interdigital beat gesture on little finger* —— 38
Figure 2.10 line 72 *high elevation upstroke on interdigital beat to ring finger* —— 39
Figure 3.1   (line 005. Defense attorney Black is on the left) —— 47
Figure 3.2   (line 009 *a diagram*) —— 48
Figure 3.3   (line 030 beat gesture upstroke) —— 48
Figure 3.4   (line 030 beat gesture downstroke) —— 49
Figure 3.5   (line 003 *go out*) —— 53
Figure 3.6   (line 005 *stays*) —— 54
Figure 3.7   (line 006 *as far as we can determine*) —— 55
Figure 3.8   (line 008 *in*) —— 57
Figure 3.9   (line 013 *through*) —— 60
Figure 3.10 (line 015 *into*) —— 61
Figure 4.1   line 07 (vertical open palm beat upstroke on *What*) —— 72
Figure 4.2   line 07 (vertical/oblique open palm beat downstroke on *people*) —— 72
Figure 4.3   line 14 (*whether*) —— 74
Figure 4.4   line 15 (lateral beat on *high school*) —— 75
Figure 4.5   line 17 (lateral beat with gaze shift on *law school*) —— 75
Figure 4.6   line 26 (high beat upstroke on *throughout*) —— 77
Figure 4.7   line 26 (beat downstroke on *treatment*) —— 77
Figure 4.8   line 33 (low arc lateral beat on *another*) —— 80
Figure 4.9   line 33 (low arc lateral beat on *widow*) —— 80
Figure 4.10 Example 2, line 12 (hand gesture on *overwhelming wave*) —— 84
Figure 4.11 Example 3 line 14 (sadness expression on *killed*) —— 85
Figure 4.12 Example 3 line 22 (sadness expression on *yeah*) —— 86
Figure 4.13 Example 4 line 7 (*empty stare gaze away from Black in 11.1 second pause*) —— 87
Figure 4.14 Example 4 line 8 (closed eyes just prior to *I think*) —— 87
Figure 5.1   line 01 (*happens*) —— 94
Figure 5.2   line 02 (*calls*) —— 94
Figure 5.3   line 06 (*voluntarily*) —— 95
Figure 5.4   line 09 (*photographs*) —— 96

Figure 5.5	line 11 (*everything*) —— **96**	
Figure 5.6	Example 3 line 01 (*microscope*) —— **101**	
Figure 5.7	Example 4 line 01 downstroke (*of the*) —— **101**	
Figure 5.8	Example 4 line 01 upstroke (*weave*) —— **101**	
Figure 5.9	Example 3 line 02 (*grass stain*) —— **101**	
Figure 5.10	Example 2 line 01 (*examination*) —— **102**	
Figure 5.11	Example 2 line 06 (*mud*) —— **103**	
Figure 5.12	Example 2 line 07 (*chips*) —— **103**	
Figure 5.13	Example 4 line 05 upstroke (*particles*) —— **104**	
Figure 5.14	Example 4 line 06 downstroke (*mud*) —— **104**	
Figure 5.15	Example 2 line 08 (*stones*) —— **105**	
Figure 5.16	Example 2 line 09 (*no nothing*) —— **105**	
Figure 5.17	Example 2 line 10 (*N'yet this dress*) —— **106**	
Figure 6.1	line 10 (***No::::***) —— **117**	
Figure 6.2	line 15 (upstroke on *to*) —— **119**	
Figure 6.3	line 15 (downstroke on *help*) —— **119**	
Figure 6.4	line 04 (*left arm*) —— **122**	
Figure 6.5	line 06 (*trapped between*) —— **123**	
Figure 6.6	line 08 (*press his chest down*) —— **124**	
Figure 6.7	line 09 (*if he raises up*) —— **125**	
Figure 6.8	lines 21–22 (*pushing down on her chest*) —— **127**	
Figure 6.9	line 22 (*to keep*) —— **127**	
Figure 6.10	line 23 (*unless he has three or four arms*) —— **128**	
Figure 6.11	line 24 (*this act*) —— **128**	

1 Introduction

Since the groundbreaking works of Atkinson and Drew (1979) and O'Barr (1981) the field of language and law or forensic linguistics has developed at a brisk and productive pace to become a dynamic fixture on the sociolinguistic landscape. From trials to jury deliberations, from credit card disclosures to law school socialization, from police interviews to citizen's emergency calls, verbal and written forms of language represent the central vehicle through which the business of law is transacted (Hobbs 2008). Rather than being the passive or neutral instrument for the imposition of legal variables, verbal and written modalities constitute the interactional vehicle through which evidence, statutes, and identity are forged into legal significance. Historically, the relationship between language and law can be traced back to ancient Rome, where lawyers were referred to as "orators" (Friedman 1977: 21), while in medieval England barristers were called "storytellers" (from the Latin "narrators," Simpson 1988: 149), both terms highlighting the legal value placed on verbal skills. More recently modern linguists define the legal order as a "law of words" (Tiersma 1999: 4) and an "overwhelmingly linguistic institution" (Gibbons 2003: 1). More humorously (perhaps), Michael Silverstein defined the law as "practicing linguistics without a license" at a Law and Society Conference some years ago.[1]

However, there is more to language and law than just written and verbal forms. Although much less studied, the integration of gesture (and other modalities such as gaze, motion, and material objects) with speech plays a significant role in legal action, one whose scholarly interest also goes back centuries. Writing on legal and forensic oratory (or "delivery" as he called it) in late antiquity, Quintilian (2001) discussed how movements of the hands synchronized with speech to convey powerful meanings – emotions that "move" or persuade the judge. He was the first scholar to consider systematically the integration of language and gesture in oratorical performance: ". . . all emotions inevitably languish unless they are kindled into flame by voice, face and the bearing of virtually the whole body . . . since words are very powerful by themselves and voice adds its own contribution to content, and gestures and movements have a meaning, then when they all come together the result must be perfection" (Quintilian 2001, Book 11.3: 87–89). Centuries later, Bulwer ([1644] 2003) examined what he referred to as the "dialects of the fingers" and its role in rhetorical practice, for instance, how lawyers use the finger of one hand to count off arguments on the other hand (a process we examine in some detail in the forthcoming chapters). In his classic study on the history of gesture, Kendon (2004: 20; see also

---

[1] And we honor his memory by using his definition as the title of this volume.

Hibbitts 1995) mentioned, "during the Middle Ages considerable attention was paid to bodily comportment and gesture for formalized gestural actions were of great importance in legal ritual."

In the ensuing chapters, we contribute to this historical trajectory and examine the complex interplay among gesture, speech and other modal resources (such as gaze, facial expression, motion, material objects) actors bring to bear on the performance of sociolegal ritual. We demonstrate in concrete detail how multimodal oratory in the law not only possesses historical significance but contemporary relevance as well. We show the crucial importance of this underdeveloped field in forensic linguistics by examining several neglected dimensions in the study of law more generally, such as objections, collective memory, bodily quotes or enactments, event structure, and multimodal poetics.

This opening chapter reviews the study of multimodal conduct in legal interaction, beginning with the trial as a distinct speech event or what conversation analysts refer to as a speech exchange system. Next, we cover gestures and other modal resources legal actors bring to bear on the performance of law, followed by the poetic function and affect in trial oratory. We conclude with a discussion of methodology and an outline of the specific chapters. Our brief overview is not comprehensive but limited primarily to the content of the ensuing chapters.

1.1 A distinct speech-exchange system?

Courtroom discourse operates in a distinct speech exchange system or speech event with differential participation rights based on institutional identity, a system of turn and turn-type preallocation (Atkinson and Drew 1979). In contrast to the local management system of everyday conversation, attorneys ask questions while witnesses answer, and this discursive division of labor occurs not only through observable patterns of interaction but in participant orientation to violations of normative context (or deviant cases), as illustrated in the following examples.

Example 1 (DA=Defense Attorney; W=Witness; J=Judge) (from Matoesian 1993)

DA: So you were aware of the meaning of this song, "Going to go downtown," and in the colloquial . . . "Gonna go and get me some ass" is that correct?
W: Can I ask you a question
J: No

Example 2 (DA=Defense Attorney; AM=Witness) (from Matoesian and Gilbert 2018)

```
01  DA:   You knew that you were going to be asked
02        questions only twelve hours ago. It was only
          twelve hours isn't that right?
03                          (1.1)
04  AM:   Yes
05                          (7.3)
06  AM:   ((slight head tilt forward and back at 7.0))
07  AM:   ((lip smack/alveolar click and head movement forward
              toward microphone with thinking face display at 7.3))
08  AM:   I would like to complete my answer on uh:: the question
09        (.) about (.) saying that Senat[or Kennedy was watching
                                         [((gaze moves to DA))
10                          (0.9)
11  AM:   ((raised and sustained eyebrow flash with open mouth
              co-occurring with three micro vertical head nods))
12                          (3.3)
13  DA:   Uh::: (.) which question are you answering now:: miss::
                                                                  [
14  AM:                                        You
15        had asked me uh::: (.) yesterday (0.5) n'you also asked
16        me this morning (0.5) **about** my statement to the police=
                          [                ]
17  DA:                   You want to-
18  AM:   =saying that *she* told me that Senator Kennedy was
19        watching. *I would like to complete that answer for the jury*
          *please.*
20                          (0.9)
21  DA:   *You mean* this is an answer that I asked you yesterday you
22        now after thinking about it overnight want to complete the
23        ah::: answer
               [
24  AM:        Ah::: No. I didn't have the opportunity to *answer your*
25        *question yesterday* ((staccato delivery))  I
26        [believe=
          [((slight gaze movement from DA to Judge and back))
27        =we had stopped [at that point
                          [((45 degree turn to Judge and back to DA))
28                          (1.2)
29  DA:   I'm sorry I thought you had uh:: completed your answer
30        If you want to say something to the jury that you've had
31        time to think about (.) [please go ahead ((markedly lower volume in lines 29–31))
                                  [
32  AM:                           No- it- >***I haven't had time to***
33        ***think about it. I would have said the same thing***
```

```
34      yesterday when you asked me.< ((Sped up))
        ((gazing at Black from 32–34))
35                      (1.4)
        [((torques upper body, tilts head/shoulders, and
        shifts gaze to thinking face/middle distance
        display in 35–41))
36      When Patty:: (1.6) came over to my house (.) when
37      she was at my house (0.4) she was sitting on my
38      couch (.) in a state of hysteria (1.6) a::::nd I had
39      asked her (.) uh few questions (1.7) Uh:: she repeated
40      (2.5) th- that he was watching, he was watching
41      (2.8) I (.) then in return asked her (.) who
42      was watching (0.8) [Was Senator Kennedy watching?
                           [((eyebrow flash))
43                      (2.0)
        [((realigns body and gaze to Black))
```

In example 1, the judge not the questioning attorney constrains the witness's verbal contribution by rejecting her request to ask the defense attorney a question. Example 2 is more complex. Notice first how the witness (AM) waits till a lengthy pause in the defense attorney's talk before *requesting* permission to complete an answer from the prior day's questioning. She does not merely produce the statement but formally requests permission, maintaining institutional and discursive identities of attorney/questioner and witness/answerer in the process. Second, after the defense attorney fails to respond (notice the 0.9 second pause in line 10), she produces a series of visual increments consisting of a raised eyebrow flash and three micro head nods in an escalating attempt to elicit a response while still withholding continuation of her projected contribution. Third, when the defense attorney starts to talk, after a lengthy 3.3 second pause, his response is prefaced with two further delay components, the first a prolongation on utterance initial *Uh:::*, the second a short micro pause – both marking violation of institutional context through perturbations in delivery of his turn-in-progress. Last, when he finally responds he neither grants the request nor gives an answer but provides a question that reformulates her request as an answer to some prior question (line 13), thus maintaining the discursive division of labor in which attorneys ask questions while witnesses answer them. In this short exchange, we can see the interactive dynamics between participants, how both shape the organization of deviant turn incursion, mark it as a departure from institutional structure through multimodal action and, in so doing, *talk the law into being* (Heritage 1984).

However, courtroom discourse is not as pre-structured as commonly assumed. First, examination in the adversary system is not organized around questions and answers but objection mediated forms of participation (the topic of chapter 2). That is, each attorney question is followed not by an answer but an objection option space, an evidential contingency that may or may not materialize during the course of testimony. Only if the attorney's question clears this option space will question-answer materialize, as in the following example.

Example 3 (slightly edited). (PA=Prosecuting Attorney; DA=Defense Attorney; J=Judge)

PA: . . . well are you saying that she was faking these symptoms of being distraught.
DA: Objection yer honor. It calls for an opinion.
J: Sustained

Although courtroom examination may appear as a question-answer system based on institutional identity, when one attorney produces a question, it opens up an objection opportunity space that, once activated, aligns different participants (the judge and attorneys), differential participation rights (one attorney objects and the judge rules on the objection), and different speech activities (grounds for the objection, warrant for posing the question, ruling etc.); and the outcome of that process affects the nature and logic of the ensuing testimony. That is, when an objection is sustained the relevance of answerer's contribution is terminated and the witness does not respond to the question – at least not verbally.

Second, that institutional structure organizes attorney-questioner and witness-answerer forms of participation does not necessarily mean that attorneys will only ask questions and witnesses only answer. Consider the following example (from Matoesian 2001).

Example 4 (slightly edited) (PA=Prosecuting Attorney; WS=Witness)

PA: And you don't have any idea how she got those bruises?
WS: If you're asking me how somebody might get bruises who's on a blood thinner I can give you any number of different reasons. She may have gotten them dancing. She may have gotten them chasing around a child . . .

In the above example, the PA's question creates the inference that the defendant caused the victim's bruises after the sexual assault. In response, the defendant poses an *if*-conditional attributed to the PA, which he then proceeds to answer in hypothetical detail. However, the PA never asked the recalibrated question. That is to say, the witness transforms questioner and answerer roles, projecting his own question to which the answer can arguably be heard, creating a subtle shift in topic. Just because attorneys inhabit the institutionally endowed role of asking questions does not mean that witnesses only answer. While the attorney asks questions, witnesses may first project "virtual" questions that displace the attorney's question, and second provide answers that align with their virtual question – fostering the impression that their *response is a reply to the attorney's question.*

And third, the possibility of appellate review in the adversary system requires the court reporter to capture the proceedings for the record (as we demonstrate in chapter 2), and problems in that task, such as simultaneous talk, introduce another contingency of participation, as in the following.

Example 5 (PA=Prosecuting Attorney; WS=Witness; CR=Court Reporter; J=Judge)

PA:	What's the it?
WS:	Patricia Bowman's account of the events that night
	is a damnable lie
	[ ]
PA:	So you're saying that she's a liar=
CR:	=I can't- I- I can't report over (( ))
J:	Missuz Lasch (.) there's been no objection from Mister Black
	but there is one from Mister Crane . . . And Mister Crane speak
	up if you can't report.

In sum, the notion of speech exchange system provides a fertile contribution to the study of language and the law because it captures structural features of legal discourse in a systematic model, a rich analytic stimulus for generating a cumulative body of replicable research findings. More problematically, however, if institutional context is much more contingent – much more improvisational – along the lines indicated and produced on a moment-by-moment basis in the unfolding stream of discursive activity – if the institution is, according to conversation analysis, *talked into being* (Heritage 1984) – then the speech exchange concept or any similar transcendent structure may well turn out to be redundant in an explanatory sense.

1.2 Multimodal conduct

1.2.1 Gesture

Multimodal conduct – the integration of speech and gesture along with embodied resources like gaze, material objects, and facial expression – represents a neglected aspect of language and law studies. Despite being omnipresent in the law, embodied conduct has been relatively ignored in contemporary research in forensic linguistics, with research being preoccupied with verbal and written modes of communication. Yet erasing this crucial feature of law leaves researchers with an impoverished understanding of the construction of legal reality. In this section of the chapter, we explore the relevance of gesture, gaze, facial expression, motion/posture and material objects for the analysis of legal oratory and how embodied resources interact with speech to reveal a more comprehensive dimension of meaning in the law.

In contemporary gesture research, McNeill (1992, 2005) defines gestures as idiosyncratic or ad hoc hand movements that co-occur with speech in the production of meaning. That is, the meaning of gesture is context-sensitive and emerges in the moment of speaking. His definition omits emblems or quotable gestures that can be understood without speech such as the "OK" gesture as well as ASL and homesign.[2] Revealing information in a different representational format, gesture functions as co-expressive with speech rather than being redundant; both convey meaning as a composite signal. Gesture employs a visual or imagistic representational format in concert with speech to encode additional aspects of meaning not expressed as effectively or economically in language. To illustrate, Goldin-Meadow (2003: 3) notes, "When a child utters 'chair' while pointing at the chair, the word labels and classifies, but does not locate, the object. Pointing, in contrast, illustrates where the object is but not what the object is." Indeed, speech does not encode visua-spatial content as effectively/economically as gesture.[3]

In contrast to speech, gesture imparts information in global, synthetic, and ad hoc representations. It lacks language-like properties such as hierarchy, segmentation, and conventional fixed form-meaning relations. Unlike language, where the whole emerges from the linear sequence of parts such as words, morphemes, phrases, etc (or in McNeill's terms analytic, conventional and combinatorial) gestural meaning of the part only emerges from the whole, where distinct meanings converge in a single synthetic gesture. Together, speech and gesture produce a richer tapestry of meaning than either could convey alone. More theoretically, gesture and speech constitute a single integrated system or two sides of the same language coin.

McNeill (2006: 60) classifies gestures as the "iconic, metaphoric, deictic and beat quartet". Imagistic gestures – iconic and metaphoric – represent some object, action

---

[2] In contrast to McNeill, Kendon (2004) finds no categorical difference between spontaneous co-speech gestures and more conventional signs, an issue relevant to gesture change from ad hoc to recurrent and to conventional signs like emblems. For example, the cyclic movement of the hand or hand rotation gesture encodes duration or aspect, a recurrent gesture exhibiting some degree of convention (see Harrison 2018).

[3] As Goldman-Meadow (2014: 73) mentions: "because the representation format underlying gestures are mimetic and analog rather than discrete, gesture permits speakers to represent ideas that lend themselves to these format, shapes, size, and spatial relationships that may not be easily encoded in speech."

or movement. That is, they convey an image or encode propositional information. Iconic gestures "present images of concrete entities or bear a close relationship to the semantic content of speech." As he puts it, "appearing to grasp and bend back something while saying 'and he bends it way back'." Metaphoric gestures present an abstract image in which a concrete source domain is mapped onto an abstract target. For example, tossing the hands to the side and behind signals a "brush-off" metaphor in which "small annoying objects" get brushed aside in order to remove them from the immediate area. Here a source domain is mapped onto the target of chasing away annoying people or insignificant ideas (Tebendorf 2014: 1549). By contrast, non-imagistic gestures consist of pointing (deictic gestures that locate referents through pointing) and rhythmic movements (beat gestures that highlight significant information without producing semantic content) in concert with speech. However, Matoesian and Gilbert (2018) found that beats may convey semantic information, and in this volume, we show how these gestures are more complex than McNeill indicates and may, on certain occasions, encode a degree of what we refer to as *residual semanticity*. Put more forcefully, the ensuing chapters demonstrate how the gestural beat is just as improvisationally dynamic and intricate as its quartet companions.

Speech and gesture synchronize for communicative effectiveness, a process that operates in and through the gesture phrase consisting of several phases, one obligatory, the others optional: preparation, obligatory stroke or meaning-bearing moment of the gesture that accompanies its lexical affiliate, post-stroke hold (often for emphasis) and retraction or return to home position. Kendon (2004) refers to this co-temporal relationship or synchronicity as the "speech-gesture ensemble" in which both modalities continuously adjust and readjust their timing so that the gesture stroke coordinates with its speech counterpart (or vice versa) to produce coherent courses of improvisational action.

As an example of their temporal contiguity, when the trajectory of an utterance is suspended or altered, so too is the co-occurring gesture. In the data segment below and accompanying images (see Figures 1.1–1.5 below from Matoesian and Gilbert 2018), the speaker's beat downstroke is temporarily suspended (line 05) and transformed in a sweeping to the side gesture to manage the new directional coordinates on the parenthetical (line 06), and when speech progressivity resumes its original trajectory, the beat recalibrates and continues its downstroke movement. More explicitly, the speaker aborts the beat downstroke while leaving the left finger suspended and then resumes the temporally suspended beat downstoke upon completion of the parenthetical (and the right hand gesture to the side). Here we see how the integration of gesture and speech functions as a finely coordinated composite signal.

*vc=parallel vertical chop
ig=interdigital gesture (tip of right index finger lands on and ascends up the fingers of the left hand)
^ig=intradigital gesture (right hand index finger latched onto the little finger of the left hand where rhythm is beat out with latching motion)

Example 6

01   She's *running* across the lawn at full tilt (0.6)
             [((vc*))   [((vc))
02   She's *being chased* by a man (0.5)
             [((vc))   [((two open palms in front of body))
03   *who is* six foot two and uh half (0.8)
             [((ig on little finger))
04   a *hundred and ninety- five   pounds* (0.9)
             [((^ig))   [((^ig)) [((^ig)) [((^ig))
             ((^IG on index finger still latched on little finger))
05   *who has-*
             [((downstroke to unfurled ring finger abort))
06   as Miss Lasch   brought out
             [((right hand lateral gesture))   [((ig onset on ring finger))
07   size *eleven and a half shoe* (0.9)
             [((ig downstroke on ring finger))
08   *who tackles* her (0.6)
             [((ig on middle finger))

Figure 1.1: *who* interdigital upstroke.

Figure 1.2: *is* interdigital downstroke.

Figure 1.3: *a hundred and ninety-five pounds* intradigital x4.

Figure 1.4: *as Miss Lasch* parenthetical or PP.

Figure 1.5: *eleven and a half shoe* interdigital downstroke.

1.2.2 Facial expressions, gaze, movement, and material objects

Facial expressions like eyebrow flash, frowns, thinking face, and disgust face (sometimes referred to as "stank face" in which the upper eyelids and cheeks are raised, nose wrinkled, and lines show up above and below the lower eyelid) signal affect, stance, and intensity of involvement in interaction. As we will see in chapter 4, closing the eyes or a blank stare may convey a somber mood. Most interestingly, facial expressions (like rolling the eyes while turning the head and opening the mouth) may convey meanings prohibited in the verbal modality, such as the attorney commenting on the witness's testimony or even the witness commenting on the attorney's question (see Matoesian and Gilbert 2018: 145–147). In chapter 2, we demonstrate how the witness recruits facial expressions as an evaluative stance on the attorney's objected to question, an act that would be prohibited in the verbal modality.

Gaze and gaze shifts – the direction of the head and eyes – co-occur with gesture, speech and posture – (the positioning and orientation of the body) – to regulate interaction; they select the recipient of an utterance, create a focus of joint attention, and contextualize emergent forms of participation in the unfolding interaction.[4]

In concert with gaze and facial expressions, various types of motion/movements and posture (such as body torque) work together in, for example, enactments to convey more vivid representations of evidence and testimony (as we will show in

---

4 In example 2 (see Matoesian and Gilbert 2018: 98–105), the witness shifts gaze from the defense attorney to a middle distance gaze (staring at an empty space to the side) and torques the upper body (or body torque) to deliver her statement (post line 34). In so doing, she confirms the attorney's metapragmatic frame that she is speaking from a prepared script.

chapter 6). They also function to display different forms of involvement and engagement in the interaction.

Finally, materiality or material conduct includes the role of objects like exhibits and how they are made meaningful at specific moments in particular contexts. Rather than seeing material objects as passive and stable features of the environment, we consider their dynamic role in constructing distinct evidential strategies. For example, in chapter 3 we show how the attorney employs tracing gestures on the surface of a diagram exhibit to attack the credibility of the witness/victim and transform opening statement into an opening argument.

In sum, a multimodal analysis studies language use not as an isolated modality but as co-occurring with other semiotic resources in the construction of legal meaning. Limiting the scope of inquiry to one modality loses the phenomenon and bleaches interactive particulars that participants orient to in interaction.

1.3 Poetics and affect in legal oratory

In his classic "Linguistics and Poetics," Roman Jakobson (1960) defined the *poetic function* of language (and differentiated it from the referential, emotive, metalingual, conative, and phatic functions) as a "focus on the message for its own sake" (1960), as "verbal art" and used the political slogan "I like Ike" as a poignant example of language use as an aesthetic performance.[5] For Jakobson, poetic features (such as parallelism, repetition, etc) foreground language form over content and draw attention to oratorical strategies rooted in sociocultural performance of identities, participation, and intertextuality. Years later, his student Michael Silverstein, added that the poetic function applies to any form of rhythmic recurrence or cardinal arrangement of equi-valued signs, not just phonological measure or prosodic structure and found that text-metrical forms signal specific interpretive strategies in both narrative and co-present interaction. That is, similarity-driven metricalized text tells listeners how to interpret and evaluate the recurrent elements. They channel and naturalize inferences through rhythmic cohesion in discursive practice. In sum, measured repetition in the aesthetic text functions as a metrapragmatic guide to calibrate interpretation and evaluation and creates not only a sense of cohesion but an affective sense of involvement in the message as well.

---

[5] As he notes in explicit detail (Jakobson 1960), the slogan consists of three monosyllables and diphthongs, with each diphthong followed by a consonant to build a vibrant and memorable flourish of sound, rhyme, and alliteration.

From a conversation analytic perspective, Atkinson (1984) and Heritage and Greatbatch (1986) investigated political speeches and how certain rhetorical devices evoked favorable responses to the message. Charismatic speakers mobilized lists of three, contrasts, puzzle/solution and combinations of these to build dramatic intensity in their speech, and when packaged in such poetic formats the message was much more likely to elicit applause from the audience than an ordinary utterance. In the courtroom, such devices allow attorneys to evaluate testimony, emphasize evidence, and cumulate inconsistencies without overtly commenting on them (which would be prohibited during questioning of witnesses), as in the following example (edited from Matoesian 1993).

Example 7 DA=Defense Attorney; V=Victim

DA: How did you wind up in his automobile?
V: I got in.
DA: **WHY::::** (increased loudness and stress)
V: Because he said we were goin to uh party at a friend of his house
DA: But you didn't know his last name, where he worked, or where he was from correct?
V: Yes.

Yet there is no *a priori* reason to limit the poetic function to language or verbal artistry. We show how the integration of metricalized gestures (like beats) with speech creates dense clusters of poetic harmony that evoke powerful emotions, a coordinated and co-improvisational ensemble designed to enhance the effectiveness of oratory and persuade the jury. Such perceptually salient and emotionally driven patterns of recurrence generate a subliminal sense of *naturalness* in and through that patterning, especially when aesthetic speech synchronizes with repetitive gestures. Indeed, cross-rhythmic patterns add a heightened degree of affective intensity to aesthetic oratory, maintaining emotional depth in the presentation and evaluation of evidence, testimony, and legal argument. We refer to this rhythmically balanced choreography of gesture, gaze, and talk as multimodal poetics or cross-modal integration of gesture and speech, and in chapter 5 we show how the defense attorney's virtuosic finger work – a dialect of the fingers – in concert with speech functions in closing argument to not only highlight the message but convey semantic information also.

The ensuing chapters analyze how the polyrhythmic interplay between speech and gesture – cross-modal metrical rhythms – modulates the affective intensity of trial oratory and dramatizes the meaning of evidence, much more so than speech alone. We demonstrate how actors establish rhythmic cross-currents to shape the presentation of evidence and place a trance-inducing spell on the audience in legal ritual. Recapturing the impulse of the classic studies on the role of speech and gesture in legal oratory, we show how the trial is not about truth or falsity but winning and losing and that depends on which side is best able to persuade the jury in and through multimodal conduct.

1.4 Data and methodology

1.4.1 We use audio-video recordings (and transcriptions of those recordings) of the William Kennedy Smith rape trial as data, one of the most infamous and widely publicized trials of the last century, involving a member of the Kennedy family and the daughter of a wealthy industrialist in West Palm Beach, Florida. After giving him a ride home from the trendy *Au Bar* nightclub, Patricia Bowman claimed that on March 30th, 1991 at around 3am Smith raped her on the lawn of the Kennedy estate. The trial took place in November-December 1991 and pitted charismatic defense attorney, Roy Black, against prosecuting attorney Moira Lasch. After a brief deliberation, the jury acquitted Smith of 2nd degree sexual assault.

1.4.2 But while we use the sexual assault trial as data, our unit of analysis is neither rape nor the rape trial or even trials more generally but the multimodal resources legal actors bring to bear on the co-construction of legal context and the co-improvisational interplay of such resources for sociocultural performance.[6] More theoretically, given our interest in aesthetic oratory we find it fruitful to conceptualize our analysis around the notion of *legal ritual*. According to Stasch (2011) ritual refers to a "poetically dense" oratory that represents group identity and generates collective sentiments, a similarity driven aesthetic performance linking macro collectivities to the micro order. Similarly, Wilce (2017: 230; see also Wilce 2009) finds that poetic devices like parallelism mark speech as ritual oratory. Wilce 2017: 81)]. Along a more critical vein, we explore how linguistic ideologies (such as inconsistency) circulate in the situated details of embodied practice to appear "naturalized" in and as mere legal reasoning.[7] In the pages that follow, we demonstrate in concrete detail how

---

[6] This leads us to anticipate a similar issue: that our data is "old" and therefore how can it be relevant several decades later. However, this confuses the case with data, conflating the two. It makes no difference how old the data is but whether or not we say something novel and interesting about the data – advancing the state-of-the-art in the field. It is not clear what it means to say that words, utterances, and gesture are "old" or "new" or "in-between."

[7] The concept of linguistic ideology stems, of course, from the work of Michael Silverstein and refers to taken-for-granted beliefs about the way language works. In chapters 3, 5 and 6 we explore how linguistic ideologies (including gestures) are interwoven into the multimodal infrastructure of legal communication.

the integration of speech, gesture and other modalities build forms of sociocultural action in ritual oratory.

1.4.3 The trial was filmed in its entirety and provided researchers unprecedented access to visual data that was previously available only on audio-tape and/or written transcripts. For the researcher, audio-visual data does not rely on memory, intuition or written-up field notes (and interpretations of ethnographic field researchers), nor does it depend on the court reporter's determination of what to include in the official transcript.[8] Researchers can view such data repeatedly to capture the delicate nuances of real-time interaction and thereby subject it to detailed scrutiny and systematic analysis, not just talk but also gesture, gaze, objects, movements, body orientation and other fine-grained features of the setting that might otherwise escape notice. Audio-video recordings of the trial, together with data transcripts, allow other researchers to check the validity of ones claims and possibly offer alternative explanations – the bedrock of scientific replication and falsification. Other researchers have access to the same data to falsify, modify or accept the analysis (rather than having to rely on experiments, ethnographic outputs and interviews) prior to the data being domesticated into *a priori* coding schemes and statistical design. Data can be stored, so that possible relevancies that may have been overlooked in prior analyses can be used to refine new dimensions and directions of study, often at much finer levels of granularity. Still, ethnographic data is often required for a more complete understanding of legal strategies and tactics, and we employ these when relevant.[9]

1.4.4 Returning to one of the classic scholars of gesture, we follow de Jorio's recommendations for conducting detailed analysis of gestures prior to conceptualization and theorizing, advice that constitutes a major point of departure for our analyses. He argues for the "importance of being exact in recognizing and describing the physical part of the gesture." Failure to accurately describe the gesture's physical position, he notes, is to ignore how "exactness can influence the interpretation of the gesture" (de Jorio 2000: 59). We also agree that observing and describing gestures in concrete detail – as situated action – in natural settings represents the first step prior to classification. In a more contemporary vein, we follow Streeck (2009: 5) who refers to his method as a "micro-ethnography" or naturalistic study of practices in social life: a methodology that eschews experiments, introspection, field notes, and/or interviews.

---

[8] Official court transcripts include only language and hence bleach multimodal conduct from view, as well as other prosodic and paralinguistic features of speech.
[9] Needless to say, we can only claim that speakers *attempt* to persuade the jury. The data we use does not allow us to make claims about the actual effects of multimodal conduct on the audience.

1.4.5 We consider how legal actors juggle a procession of rhythmic and harmonic choices to build a dynamic and improvisational interplay of speech and gesture in the production of evidential reality. We show their relevance and significance for understanding legal oratory. While it may seem rather "cosmic" we let the data take us where it wanted to go, rather than where we wanted to go, rather than impose our own pre-conceived ideas about where analysis should go and what it should include.

In her classic work, *Hearing Gesture*, Goldwin-Meadow (2003: 3) writes: "To ignore gesture is to ignore part of the conversation." To revamp her proposition slightly, if conversation is multimodal then it follows that the study of legal discourse needs to incorporate a more comprehensive view of the communicative process in its sociocultural context: language use as the improvisational integration of verbal and embodied modes in the contextualization of legal meaning. Ignoring this other half of the conversational equation leaves us with an incomplete understanding of law and the sociocultural dynamics embedded in it.

1.5 Chapter overview

The book not only studies the integration of speech and embodied activities in the construction of legal context. It does so by exploring a number of underdeveloped aspects of sociolegal ritual more generally.

The first part of Chapter 2 argues against the assumption that trial examination consists of question-answer patterns. Rather, data from cross-examination in a criminal trial reveals the existence of an objection option space that screens questions for evidential relevance. When activated, the option space transforms the micro social organization and participation structure of the court in officially sanctioned ways. Unofficially, however, both witness and judge deploy forms of multimodal conduct as an evaluative stance to comment on the objected-to-questions and signal egregious evidentiary violations to the non-questioning attorney and jury, undermining the credibility of the questioning attorney. The second part of the chapter analyzes an objection conference after the judge dismisses the jury to consider the questioning attorney's litany of evidentiary violations. The only effective cross-examination in the entire segment occurs when the judge questions the attorney on her blatant disregard for proper legal procedure. The judge employs speech-synchronized gestures in a poetic performance to insult the questioning attorney, a legal identity constructed in and through a multimodal mapping of denotational text to interactional function.

The next two chapters analyze opening statement (with a brief "detour" to direct and cross-examination at the end of chapter 4). Chapter 3 shows how an attorney's multimodal narrative transforms an opening statement into an argument about the credibility of the main prosecution witness. The defense attorney integrates speech, gestures, and exhibits to shape motion events and spatial images into relevant objects of evidentiary knowledge, creating inconsistencies in the witness's account under the auspices of merely showing the jury locations and movements in the defendant's home. We demonstrate how the encoding of motion events percolates in and through a polyrhythmic and multidimensional poetic format to naturalize gender ideologies – cultural expectations governing victim identity – in the social construction of rape's legal facticity.

Chapter 4 examines how collective memory and cultural trauma inhere in the multimodal interplay between macro structures of space-time and microcosmic action. We show how collective memories and cultural sentiments function in the concrete details of poetic oratory and emotional charged speech to frame evidence, construct legal identity, and shape the interpretation of testimony. Legal actors integrate language, gesture, and gaze to shift the plane of legal reality into a sacred performance, a solemn, co-operative ritual that lets lawyers and witness co-construct an emergent space for jurors to step into history as a socio-legal strategy.

Chapters 5 and 6 examine closing argument. Chapter 5 analyzes how gestures modulate the affective intensity and epistemic certainty of speech in closing argument. We analyze how gestures synchronize with speech to not only orchestrate the rhythm of utterances but also persuade jurors of the truth of one side and falsity of the other. We demonstrate how closing functions as a multimodal narrative in which gesture, gaze and speech transform legal evidence into institutionally organized forms of persuasive oratory: how attorneys foster the impression that they really believe their side should win by organizing and emphasizing favorable points of evidence, more explicitly by counting off and counting up inconsistencies in the prosecution case.

Chapter 6 investigates a semiotic modality that figures prominently in interdiscursive connections, an embodied resource that circulates between the historical and current speech event. Speakers do more than merely report the spoken or written word from historical context. They use their bodies to enact and demonstrate prior actions, actors and events in the current speech event. They use gestures and other embodied modalities to enact prior moments of legal relevance for the evidential task at hand. While a number of studies have examined the legal relevance of reported speech and written documents few if any studies have shown how gestures and other embodied modalities function in the enactment of evidential context. We examine how the attorney in closing argument does quite more than merely report historical statements; he enacts them in embodied form to provide a more captivating experience for the jury.

The final chapter provides a summary of our main findings and then explores how they relate to recent recommendations by feminist legal scholars for reforming the injustices of the rape trial.

One final note. Our use of the term "oratory" is neither merely idiosyncratic nor stylistic but reflects an interest in the art of legal persuasion – how oratorical performance is designed to move the jury to accept one version of evidential reality rather than another and to win the case. This should not be surprising. "Oratory" first appeared as a term neither in politics nor religious contexts but in the law courts of Athens centuries ago. Indeed, our interest in the multimodal tactics of effective discourse can be traced to Quintilian himself – who was originally a "pleader" in the court of law (Kennedy 2013).

2 *This is not a course in trial practice:* Multimodal participation in objections

Numerous researchers argue that institutional interaction is organized around questions and answers. According to Tracy and Robles (2009: 131), "Questioning is one of, if not *the*, central communicative practice of institutional encounters." In a similar vein, Freed and Ehrlich (2012: 3) state: "The study of questioning has always been central to investigation of institutional discourse." Speech events like interviewer-interviewee (Clayman and Heritage 2002), doctor-patient (Heritage 2010), teacher-student (McHoul 1978), police-suspect (Komter 1998), counselor-counselee (Erickson & Schultz 1981), and lawyer-witness (Atkinson & Drew 1979) consist of asymmetrical roles and discursive control that organize questioning practices in institutional settings.

Courtroom examination between lawyer and witness, in particular, consists of identity-driven question-answer patterns, as a number of scholars have indicated. According to Heritage & Clayman (2010: 176; see also Atkinson & Drew 1979), the turn taking system in trial examination is "organized around questions and answers." Raymond (2006: 115) states: "lawyer's conduct in courtroom interactions . . . is organized primarily through questions and answers." Tracy and Robles (2009: 137) remark that, "The drama of the Anglo-American legal system is all about questioning, particularly in cross-examination." Finally, even the noted forensic linguist and attorney, Peter Tiersma (1999: 168) claims: "The conversation between the lawyer and witness . . . consists of virtually nothing but questions and answers."

Indeed, perhaps no area of institutional interaction possesses more significant consequences for questioning than courtroom examination. Because lawyers control the turn taking system in court they set the agenda or topic, limit answers from witnesses, and phrase evidence to steer a particular interpretation of testimony through questioning practice. As Holt & Johnson (2010: 21) put it: "The most distinctive and widespread linguistic feature of legal talk is the question . . . lay interactants are largely controlled by and at the mercy of questions from professionals in dyadic legal encounters." The classic quote from Sacks (1988: 54) captures the controlling power of questions, especially when based on the asymmetrical distribution of institutional options: "As long as one is doing questioning, then in part they have control of the conversation."

In this chapter, we take a radically different position on questioning practices in court and the *question centric* focus characteristic of trial examination in the adversary system. We argue that direct and cross-examination are not organized around questions and answers as commonly assumed. Instead, we show how they are organized around what we refer to as *objection mediated forms of participation*.

That is, each attorney question is followed not by an answer but by an *objection option space*, an evidential contingency that may or may not materialize during the course of testimony. Only if the attorney's question *clears* this option space will question-answer and institutional control materialize. More explicitly, the question has to clear the option space after screening for evidential relevance en route to completion. Thus, rather than question-answer pairs, direct and cross-examination are organized around (1) question, (2) objection option space, and (3) answer. If the non-questioning attorney activates the objection option space, the participation structure changes from attorney-witness to attorney-attorney-judge (or attorney-judge) and from question-answer to objection-(basis for question)-ruling, a contingent and improvisational evidential field consisting of multi-laminated, multiparty, and multimodal forms of participation.

The chapter is organized as follows. After a brief literature review of objections we introduce an audio-video transcript from cross-examination in the trial. Section two analyzes a four-minute segment of data consisting of twelve questions by the prosecuting attorney, ten of which are objected to by the defense attorney, with all ten sustained by the judge. We show how the witness mobilizes body and head movements, gaze, and paralinguistic cues as an affective stance to comment on the prosecuting attorney's questions and signal egregious evidentiary violations to the defense attorney and jury. We also show how the judge uses gaze and facial expressions to move out of a neutral "umpire" role and signal objectionable questions to the non-questioning attorney. More generally, we demonstrate how the institutional reflexivity of questioning practices departs significantly from orthodox descriptions of question and answer patterns in adversarial courtrooms. The final section investigates how a series of sustained objections leads the judge to dismiss the jury and deliver a multimodal "cross-examination" within a cross-examination, a surprise mapping of denotational text to interactional function (Silverstein 1998). Both sections track the ongoing flow of multiparty participation shifts and their multimodal laminations during the course of cross-examination.

2.1 Literature review

In her study of the O.J. Simpson trial, Cotterell (2003: 95) counted more than 16,000 objections, though she failed to provide an analysis (keeping in mind that she never set out to do so). Despite their prevalence in the adversary system very few researchers analyze the organization of objections in the courtroom (but see Heffer 2005: 82–84 for a brief discussion). To be more precise, the only (detailed) discursive analysis of objections (at least to the best of our knowledge) was in Matoesian's (1993) study of rape trials thirty years ago, and it will be instructive to review briefly several of the major findings before proceeding. First, he demonstrated that the majority of objections occur in turn environments prior to the witness's answer, thereby preempting it, and even when the witness begins an answer the attorney uses turn-interruptive objections to "strand" the answer in progress. Second, objections are not a type of remedial insertion or repair sequence, since the canonical insertion sequence maintains the integrity of the question- answer pair, even across multiple embeddings. By contrast, when the judge sustains the objection it deletes the relevance of an answer altogether. And, finally, even when the judge overrules an objection, the questioning attorney typically reformulates the question, creating a "fresh" question-answer pair in the process (for numerous reasons). Consequently, it seemed necessary to build an objection option space into the very fabric of questioning practices – not questions but objection mediated questioning practices. In the case here, we build on this prior work but now with a renewed emphasis on the role of embodied conduct, interactive forms of participation, and affective stance, especially in the conduct of the witness and judge.

Although language and law research ignores objections, trial practice textbooks typically devote an entire chapter to their study, a consideration that may be instructive for linguists studying legal institutions. According to Haydock and Sonsteng (1991: 152), objections are a "procedure used to oppose the introduction of inadmissible evidence, to oppose the use of improper questions and to stop inappropriate conduct during the trial." Perrin et al (2003: 343) state, "Any overall trial strategy that fails to include consideration of objections and even objection strategies is inadequate and incomplete." And it is easy to see why. In the adversary system, each attorney question invites careful screening for its evidential relevance. Each question must be precisely phrased to exclude improper evidence and procedure during the trial – legal strictures that may be relevant "down the line" should the case go to appellate review. Just as important, sustained (or upheld) objections may undermine the credibility and authority of the attorney. As Haydock and Sonsteng (1991:153) note, objections "signal lack of ability on the attorney when sustained ... that they are poorly prepared and possess little knowledge of evidence law." Mauet's classic text (2017: 515), links objections to legal identity, in particular, which attorney occupies the role of "evidence expert": "Jurors notice who wins objections and how the judge reacts to objections and rules on them ... it shows who the better attorney is." This relates to a legal point we mentioned in the introduction. The trial is not about truth or falsity but winning and losing, and that depends on which attorney can best persuade the jury. Indeed, the credibility of the attorney is under scrutiny and evaluation just as much as the witness.

However, while trial advocacy texts devote an entire chapter (or more) to objections, they possess what we refer to as a *speaker centric* folk ideology of communication that ignores the role of the witness. In this chapter, we not only examine the objection work of attorneys and judge but also analyze the dynamic performance of the witness in the interactive and embodied co-construction of "evidence expert."

Objections: Example 1. Approximately 4 minutes. PA=Prosecuting Attorney, WS=William Smith (the defendant), DA=Defense Attorney, and J=Judge

```
01   PA:   Several people saw Miss Bowman (1.5) in the time period after
02         this took place. (1.6) Doctor Prosko saw her (0.2) and testified as to her
03         physical injuries and her emotional condition, (1.1) that she was
04         weeping, that she was withdrawn, that she was regressed. (2.2) Mr.
05         Smith are you trying to tell the jury that all these symptoms (.) resulted
06         as a reas- consequence of consensual sex between two adults?
07   WS:   I don-
                [
08   DA:        Objection yer honor. (.) It calls for an opinion.=
09   J:    =Sustained.
                            (.)
10   PA:   Well that's what you're trying to tell the jury that there was nothing
11         but consensual sex here. Isn't that your story?
                            (.)
12   WS:   It's not my story. That's the truth
13   PA:   OK. Isn't that what you're telling the jury. It's consensual?
                            (.)
14   WS:   I'm telling you that we had sex (.) and that (1.5) it was her decision and
15         my decision and we- we did it an- and that's the way it happened.
16   PA:   Yet afterwards she went to a hospital for medical attention (1.5)
17         and a physician testified that she was weeping.
                       [    (.)    [   (.)        [
18   WS:             I-      canno-        tell y-
                                              [
19   DA:                                   Objection.
20         It's argumentative yer honor.=
21   J:    =Sustained.
                            (1.0)
22   PA:   Her emotional condition (.) after this supposedly consensual sex
23         was withdrawn (.) upset (.) distraught. Are you try- trying to tell the jury
24         that these are natural reactions of consensual sex?
                                                [
     WS:                                        ((open mouth suspension till
                                                  line 27))
25   DA:   Objection. It's argumentative yer honor.=
                                         [
26   J:                                  =Sustained
27   WS:   ((lateral head shake co-occurring with audible exhale))
                            (6.0)
28   PA:   Doctor Prostko testified that consensual sex is not considered
```

29		traumatic event. You'd agree with that testimony wouldn't you?
		[
	WS:	((open mouth suspension till line 32))
30	DA:	I Object yer honor (.) It calls for his opinion.=
		[
31	J:	=Sustained.
32	WS:	((body recoil, head tilt, pursed lips and audible exhale))
		(7.1)
33	PA:	Miss Bow- well are you saying that she was faking these symptoms
34		(.) of being distraught.
		[
35	DA:	Objection yer honor. It calls for an opinion.
36	J:	Sustained.
37	PA:	Well the testimony is that she was *extremely distraught* (.) and upset
38		in the morning hours and afternoon of March Thirtieth, Nineteen Ninety-
39		One (.) Wouldn't you agree here that a little more happened here than
40		consensual sex?
41	DA:	Objection yer honor. *Same* objection. *Same* line of questioning.
		[
42	J:	Sustained.
		(.)
43	PA:	Well how do you explain (.) *Mister Smith* the fact that she was
		extremely distraught, crying, that she went to a *rape* crisis line (.)
		that she sought medical attention=
		[[
46	WS:	((open mouth)) ((head shake with co-occurring audible exhale,
		closed eyes and upper body recoil))
47	DA:	=We object and may we approach?
		((transcription doubt, J's sustain?))
		((Side Bar Conference for Over Five Minutes; inaudible)) ((Return
		to Cross-Examination of Defendant after Sidebar))
48	PA:	Wouldn't you agree Mister Smith that Doctor Prosko is a more
49		qualified individual than yourself (.) to determine whether an
50		individual was traumatized
51	DA:	Objection yer honor. Calls for an opinion.
52	J:	Sustained.
		(.)
53	PA:	Didn't you find her a very competent, caring physician when she
54		testified in court (.) in front of the jury.
		[
55	DA:	Objection yer honor.
56	J:	Uh::I sustain the objection. Questions concerning this witness's opinions
57		(.) regarding the qualifications or opinion testimony of other witnesses are
58		clearly improper (.) and uh I think after six or seven objections being
59		sustained I've made myself clear. (1.0)

```
60   PA:   What about detective Rigolo. She saw Mi- Miss Bowman=
                                         [
61   J:                                  ((open mouth gaze to DA))
                                         ((1.2 seconds))
62   PA:   =after the uh-
63         report to the Sheriff's Office on March, Thirtieth, Nineteen Ninety-One=
                                         [
64   J:    ((second open mouth gaze to DA for 1.1 seconds))
65   PA:   =Were you aware that she also saw her in a distraught mood.
                                         [
66   WS:                                 ((gaze to jury and audible exhale))
67   DA:   Objection yer honor. Assumes facts not in evidence and I move that the
68         question be stricken and the jury instructed to disregard.=
69   J:    =Sustained. Members of the jury, um (.) questions (.) by lawyers are not
70         evidence (.) and u::m I need to confer I think with these attorneys outside
71         of your presence for a minute or two. Why don't you step into the jury
72         room.
```

2.2 Embodied participation and affective stance

In lines 01–06, the prosecuting attorney (PA) questions the defendant (WS) about testimony from the emergency room physician, Doctor Prosko, on the victim's physical and emotional condition after the sexual assault. She then formulates a strongly accusative question based on those facts: *are you trying to tell the jury that all these symptoms resulted as a consequence of consensual sex between two adults?* In line 07, WS's answer is cut-off and "stranded" in route to completion by the defense attorney's (DA) interruptive objection (line 08), an objection based on improper opinion. That is, only the jury draws opinions and conclusions from testimony, not the witness, who may only testify to facts or his direct perception of events. And despite PA's rhythmic parallelism in the relative clauses and logical narrative that embeds a deftly crafted contrast between symptoms and consensual sex (lines 01–06), the judge (J) immediately sustains the objection in line 09.

Several empirical observations support our prior points about objections. First, DA's objection preempts WS's answer in progress. WS cuts-off his utterance and relinquishes his turn immediately at objection onset, demonstrating that even for an answer-in-progress, the attorney's interruptive objection takes precedence over continuation of the answer. Second, the objection transforms participation structure from PA-WS to DA-J- PA, who now provide the basis for the question-objection (PA and DA) and judicial ruling (J). According to Goffman (1981), participation refers to the micro-organization of communicative roles in focused interaction, the institutional expectations and obligations of discourse identities and the relevant activities they may engage in on a moment-by-moment basis. In his words (Goffman 1981: 3), "When a word is spoken, all those who happen to be in perceptual range of the event will have some sort of participation status relative to it. The codification of these various positions and the *normative specification of appropriate conduct* [our emphasis] within each provide an essential background for interaction analysis . . ." Third, and with these points in mind, while PA may provide the grounds for her question after DA's objection (and she posses the institutionally endowed right to do so), J's immediately latched sustain in line 09 forecloses any possibility of that happening; the violation is so egregious and blatant that she does not even give PA an opportunity to provide the grounds for her question. And, last, J's immediate post-objection sustain may subvert PA's evidential ability, undermining her credibility as an "expert" in evidence law.

In lines 10–14, PA poses two questions regarding consensual sex, and WS answers both he and the victim consented to the issue in question. In line 16, she mobilizes a formally marked contrast off WS's answer in line 14: *Yet afterwards she went to a hospital for medical attention and a physician testified that she was weeping.* However, DA's objection (lines 19–20) preempts WS's turn interruptive answer, which stops immediately at objection onset (on line 18). The DA objects to PA's question on the grounds that it is argumentative, or that the "lawyer is arguing his point to the jury rather than seeking information from the witness" (Perrin 2003: 363). It refers to a question where the "lawyer wants to make a speech and does not care what the answer is" (Tanford 1983:297). And such speeches belong in closing argument not cross-examination. Once again, J sustains DA's objection immediately with the latched ruling (line 21), foreclosing any possibility for PA to provide the basis for her question – suggesting to the jury that she is poorly prepared.

Although objections bring a new form of participation into play, consisting of J, PA and DA, transforming WS from a speaking to non-speaking participant, WS's response (in lines 27, 32, 46, and 66) co-occurs within the objection framework as a subterranean form of participation not officially authorized or sanctioned by the court: what we refer to as a *residual form of embodied participation.* That is, although WS withholds his answer at objection onset, he continues into a new but not competing form of participation in the objection context, one that takes an embodied and paralinguistic stance toward a target of value in the sociocultural field: toward PA's question.

While stance orthodoxly refers to the speaker's degree of commitment to and certainty of the proposition marked through morphosyntax, it also expresses emotions encoded not only through language, but through embodied conduct, such as gaze, head and body movement, and paralinguistic conduct (such as audible aspirations). As Biber et al (1999: 967) state: "emotive stance is conveyed through non-linguistic means: body posture, facial expression, gesture and paralinguistic devices." Moreover, affective stance not only encodes emotion, but also evaluates the object of that emotion. Indeed, the integration of embodied and paralinguistic resources generates a powerful affective stance, a stance that "evaluates other's claims and statuses" (Jaffe 2012: 7) and, in the case here, it evaluates PA's status as an evidence expert.

Just as important embodied conduct represents a perspicacious method for keeping multiple forms of participation in play simultaneously (affective evaluations that would not be allowed in the verbal modality. And such evaluations will never appear in the court transcript). At a finer level of granularity, even though WS stops speaking at objection onset, he still takes an affective stance through embodied conduct, subliminal rhythmic interjections organized as follows.

Although WS completes two objection free answers in lines 12 and 14, he aborts both utterances in lines 07 and 18 (after a few utterance tokens) at onset of DA's objection. After that he executes a dramatic multimodal display consisting of the following. First, after PA's questions in lines 22 and 28 he opens his mouth as if to answer but then suspends it in an open mouth display after DA's objection, while in line 46 he combines an overlapping open mouth and subsequent embodied display (over PA's question in lines 43–45; see Figure 2.1 below). In all three cases, the displays operate as the interactional platform for delivering a noticeably marked post ruling exhale or what we refer to as *exasperated aspiration* in lines 27, 32 and 46. His paralinguistic manifestation of affect encodes feeling and evaluation to signal a sigh of frustration regarding a litany of improperly phrased questions. Second, exasperated aspiration co-occurs with three lateral headshakes in line 27 and body recoil, oblique head tilt, lateral headshakes, and pursed lips in lines 32 and 46 (see Figure 2.2 below). The head tilt, in particular, represents a quite revealing method of encoding affect. According to Calbris (2016: 96–98), the "head tilt represent a point of view: the slanted, imbalanced position of an objection in relation to the vertical axis... oblique position of head mimics a physical deformed head symbolic of something negative." And here the tilt indexes the 'imbalanced' and 'deformed' questions posed by PA. WS modulates the affective intensity of evaluation through this dynamic interplay of embodied and paralinguistic signals, semiotic resources that register a sense of futility in and through an emotionally charged performance. In essence, just because WS does not answer the question does not mean he fails to respond. He uses a residual and concurrent form of embodied participation – affiliating with J and DA – to impart a damaging emotional stance towards the PA.

A poignant example of this damage occurs at the end of her question in line 43. As we have seen in line 46, WS recruits a series of head shakes co-occurring with audible exhale, closed eyes and body recoil that overlap her turn final component ("medical attention") registering a sense of bewilderment after so many improper questions. And his residual participation leads not only to an objection but to a side bar conference in which DA asks J to declare a mistrial because of a series of improper questions PA intentionally posed to WS.

Figure 2.1: line 46 *open mouth suspension.*

Figure 2.2: line 46 *body recoil, head tilt, pursed lips and closed eyes.*

And, third, after PA's questions in lines 33, 37, 48, and 53 WS neither begins an answer nor produces an open mouth display or any other form of nonverbal action. After PA unleashes a litany of improperly phrased questions, he produces a progressive and incremental de-escalation of involvement in the answer role, as if becoming so accustomed to the predictable violations that he does not even gear into a turn (of course, that lack of response is itself a response). That is, WS displays an orientation to the impropriety of PA's questions as her violations progress (a way of displaying lay expertize perhaps over the legally trained "expert!").

Consider WS's involvement in the answer role in more detail. In lines 07 and 17 he starts then aborts answer at objection onset, displaying high involvement in the answerer role. At the end of her questions in lines 24 and 29 he withholds an answer while recruiting the open mouth suspension to deliver exasperated aspiration, recoil, tilt and lateral head shakes after J's sustain, an intermediate display of involvement. But after the sidebar conference he withholds any form of response except gaze and exasperated aspiration toward the jury (in line 66), as if to say, "can you believe this" or low involvement in the answerer role. More theoretically, we see an incremental deceleration of involvement from high to intermediate to low.

At the same time, notice the spatio-temporal environments in which his verbal, embodied and paralinguistic responses occur. In lines 07 and 18, WS's verbal responses start and then stop at objection onset; in lines 27 and 32 his embodied and paralinguistic responses occur after J's sustain; in lines 46 and 66 embodied and paralinguistic responses overlap PA's turn. His recognition onset displays accelerate – incrementally and progressively – as violations progress. He decelerates verbal and non-verbal involvement as he simultaneously accelerates recognition displays of evidential violation. (from starting but aborting an answer to post sustain embodied conduct and then to question co-occurring embodied conduct).

In sum, answer involvement decelerates from high to intermediate to low – from starting then aborting, to open mouth suspension etc, and finally to verbal/ nonverbal withhold. Recognition onset timing proceeds in the opposite direction, accelerating from low to intermediate to high based on changing contextual coordinates: from post objection (aborting the answer) to post sustain and finally to question overlapping. Both involvement and recognition displays move incrementally and progressively in opposite directions as the litany progresses.

More theoretically, the interaction between involvement and recognition functions in the following way. WS lowers degree of involvement as he recognizes the violations. How does he recognize violations and activate changes? He monitors and analyzes the unfolding objections in a situated micro socialization event that furnishes the interpretive resources for understanding and predicting evidential violations, properly from improperly phrased questions, and for aligning his stance in response to other legal actors and actions: affiliating with J and DA and disaffiliating from PA.

Nor is the defendant the only one who uses verbal and visual conduct with an eye toward the evidential work it can accomplish. J mobilizes multimodal conduct in several ways, primarily by sustaining the objection. However, it is not just her sustain but the stunning series of immediate post-objection sustains – with no hesitation or pause – that questions PA's credibility and authority (lines 09, 21, 26, 31, 36, 42, 52, 56). In fact, in lines 26, 31, and 42 J's sustain overlaps DA's objection: the tag in line 26, "opinion" in 31, and "questioning" in 42. Moreover, in line 56 J not only sustains the objection but also delivers a trenchant admonishment to PA in front of the jury.

Just as impressive, in lines 61 and 64 J produces two instances of question overlapping embodied conduct. In line 60, she gazes at PA with her mouth closed (see Figure 2.3 below). However, during the address term in line 61 (over Mi- Miss Bowman), she shifts gaze with an open mouth to DA, and then returns a closed mouth gaze back to PA (see Figure 2.4 below). In line 64, she shifts to another open mouth gaze toward DA. Her shifts in gaze and facial behavior signal a transparent violation to DA – in full view of the jury – while not even waiting for question completion, anticipating an objection to facts not in evidence. Finally, in line 69 she not only sustains the objection but also criticizes PA and then dismisses the jury in a most conspicuous augury of things to come once outside their presence. Although she directs her post-ruling comment (line 69) to the jury, it functions as an indirect reprimand to PA, one that constitutes the "last straw" as it were in a long litany of sustained objections.

Figure 2.3: line 60 J closed mouth gaze at PA.

J's gaze and open mouth shifts mark a heightened state of involvement that projects an affective stance toward and negative evaluation of PA's questioning. J's sustains are latched, immediate, and overlapping, latched as in lines 09, 21, 31, and 70, immediate post-objection environment as in lines 36, 52, and 56 and overlapping as in 26 and 42. Such quick sustains show affiliation with DA's objections and disaffiliation from PA's questions, implicitly questioning her competence as an evidence expert. As mentioned previously, jurors monitor who wins objections and how the judge responds to them (Mauet 2017: 515). And given the rulings thus far PA's position looks precarious, to say the least. While DA and J oppose PA's questions through rules of evidence, WS evaluates them in and through a sociocultural integration of multimodal resources. Together, each mutually elaborates the other through emergent forms of participation to undermine PA's credibility and competence as an evidence expert.

Figure 2.4: line 61 *open mouth gaze at DA.*

More theoretically, the organization of question-answer patterns in court relies on this metapragmatic field of embodied participation that negotiates what constitutes a proper question in the first place, a contingent enactment of felicity conditions that governs questioning practices based on evidential relevance. More forcefully, participation roles take metapragmatic precedence over sequential relevance (like adjacency pairs from conversation analysis). One cannot ask any question in court anymore than one can in everyday conversation. Questions rely on the participation status of the speaker to ask the felicitous questions to the institutionally endowed interlocutor, felicity determined by an official who judges – and simultaneously becomes part of – the reflexive organization of questioning. And that reflexivity shapes and is shaped by the judges enactment of linguistic ideology. Only testimony from witnesses not questions from attorneys constitute evidence, and discursive organization in court reflects that institutional signature: how courtroom participants invoke felicity conditions as a contingent and practical evidential accomplishment (that is, an accomplishment that may or may not be realized).

In this section we have seen how participation shifts from PA-WS to DA-J, while WS shifts into a residual and multimodal lamination of affective participation not authorized by the normative order of the court. That is, embodied participation and affective displays manage communicative work prohibited in the verbal modality. In the next section, we will see the multilaminated, multimodal, and multiparty outcome of the objection participation framework after J dismisses the jury, in particular how the objection conference not only slows down the adversary proceedings but also foregrounds the identity of the prosecuting attorney and its relation to the metapragmatics of questioning practices.

30 — 2 *This is not a course in trial practice:* Multimodal participation in objections

After dismissing the jury, J invites both attorneys to argue their respective positions, beginning with PA below.

Objection Conference: Example 2 (CR=Court Reporter)

Objections: Conference Approximately 6 minutes (starts at 4:05 on tape)

```
01   PA:   Judge Lupo I'm somewhat at a loss here I cannot (0.5) understand (.)
02         in a case where the allegation's rape (.) the victim has stated (0.5)
03         that she was severely emotionally traumatized by this (0.5) We have
04         several individuals who saw her there (.) and now mister Smith is
05         giving a different version (.) . . . obviously it's appropriate in cross-
06         examination to bring out (0.5) other people I- fo- I'm going to
07         bring these people in for rebuttal of his testimony . . .
08                         (4.5)
09   J:    Mister Black you want to argue your position.
10                         (0.5)
11   DA:   Yes your honor (.) Detective Rigolo has not testified (.) For the prosecutor
12         to state in her question that detective Rigolo saw this person in a
13         particular condition is totally improper (.) It is suggesting evidence to the
14         jury that's *not* in the record (.) assuming facts not in evidence (.) I think
15         it's (.) improper *even* if she had testified it'd be improper. He can testify
16         as to what *he* says he saw and observed and heard. She can ask him
17         *any* question that she *wants* about that but she can't ask him to say
18         what other witnesses say. That's (.) improper.
19                         (6.0)
20   J:    Anything further from the state.
21                         (5.0)
22   PA:   Yes judge I'm going to ask him and it's not facts not in evidence.
23         I'm asking him if other people (0.4) saw her would that be a difference
24         (.) from his story. I mean it's bringing out impeachment of his testimony=
25   J:    =Well I rule question by question and the last question was (1.0)"What
                 [         [            [              [             [
                ((PB))   ((PB))       ((PB))         ((PB+hold))   ((PB))
26         about detective Rigolo (.)   were you aware that she also saw her in a
                 [      [             [                  [        [ ((PB))
                ((PB))               ((PB))            ((PB))   ((PB))
27         distraught mood." Now that is clearly (0.3) an improper question.
                 [       [             [                [
                ((PB))  ((PB))       ((PB+hold))     ((2 lateral PB))
           ((PB=precision ring gesture in up/down beat unless otherwise
           noted))
28                         (.)
29   PA:   Could the court explain to me why?
30                         (2.3)
31   J:    ((turns to court reporter=CR)) Would you read bla- back mister
```

32		Black's (1.09) uh- uh- basis for his objection
		((CR is typing on a steno machine to record the proceedings))
33	CR:	((15.2 seconds as CR scrolls up steno paper for basis))
34	CR:	((stops scolling))
35	CR:	(turns to J)) The basis for his objection?
36		(6.5)
37	J:	Let me repeat myself OK. The question that Miss Lasch-
38		[[
39	CR:	((Scolls back down steno paper)) ((raises finger for J to pause))
40	J:	((waits for CR to get ready to transcribe 4.5 sec))
41	J:	The question last proposed to this witness by Miss Lasch (0.3) 42
		goes (1.4) somewhat as follows: *"what about detective Rigolo*
43		*(.) were you aware that she also saw her in a distraught mood."*
44		(.)
45	J:	((to PA)) Do you agree with mister Black that detective Rigolo
46		has not testified?
47		(.)
48	PA:	She will be testifying on rebuttal
49		[
50	J:	***Do you agree with mister Lasch that***
51		***she has not tes****- Mister Black that she has not testified.
		[ ]
52		*((shakes head w/closed eyes over repair & motions with*
		*open palm to DA)) ((**italics in bold=loud+staccato***
		in lines 50, 55, & 74))
53	PA:	No she has not testified but (( ))
54		[
55	J:	***Then how in heaven's name is that a***
		[ ]
		((lateral head shakes))
56		***proper question***?
57		
58	PA:	Because it's an entire inconsistency with what his testimony is . . .
59		it's impeachment . . . He can be impeached by detective Rigolo's
60		testimony . . .
61		(1.0)
62	J:	Miss Lasch (10.8) this is not a course in *trial practice* (1.2)
		[((closed eye lateral head shakes during 10.8 sec pause))
63		Of course you can impeach by bringing in detective Rigolo.
64		(1.0) But you *cannot* properly ask (.) this witness *"were you aware*
65		*that she also saw her in a distraught mood."* (4.8) What about
66		detective Rigolo prior to her testifying (.) Now I'm not going to
67		go through (0.9) a course (1.4) and I- and I- just have to believe (.) that
		[ ]
		((closed eye head shakes))

68	(0.5) these are um:: (1.7) questions being intentionally proposed
	[ ]
	((tilts head+oblique head shakes))
69	to this witness that are clearly improper (.) Either that or um:::
	[ ]
	((oblique lateral head shakes))
70	(.) I'm very concerned . . . Now this witness will not be asked
71	his opinion testimony on the opinion testimony of other (0.6) experts [((interdigital beat upstroke on "his", downstroke/hit on first "opinion testimony", and push down on "other")) ((beat on little finger of LH with index finger of RH; hold during 0.6 sec)) ((RH=Right Hand; LH=Left Hand)) (2.9)
72	He will not be asked about (0.8) facts that are not in evidence (5.2)
	[((interdigital beat upstroke, downstroke and push down] ((beat upstroke on "He" suspend 3 sec with very high elevation ;
	beat lands on ring finger on "facts"; pushed down on "evidence"))
73	And (0.7)
74	you are free to cross examine him as you would **any other witness** (.)
	[((intradigital beats]
	((B*)) ((B*)) ((B*)) ((B*)) ((B*)) ((B*)) ((B*)) ((B*))
	(("**any other witness**"=staccato. 8 intradigital beats on middle finger; final 3 staccato on "**any other witness**")) B*=Intradigital Beat where beat index finger stays on middle finger of the other hand while beating))
75	Bring out the jury please.

2.3 Objection conference: Cross-examination within cross-examination

Objections not only function retrospectively to reflexively problematize questions but also prospectively towards the imminent procedural liabilities for engaging in inappropriate conduct. In this regard, the center of gravity shifts from the role of improper questioning in the objection sequence to a metapragmatic attack on credibility and professional identity in the objection conference. Once the jury departs participation structure changes from PA-WS-DA-J to J-PA-DA (at least for now), with WS in a co-present but non-participating role. Although J never gives PA an opportunity to provide the basis for her objected-to-question during cross-examination, she gives her two opportunities in an objection conference. In lines 01–07, PA proceeds to do precisely that. Her basis consists of a more or less general claim about the right to bring in rebuttal witnesses to impeach WS's testimony. Notice that her *be going* construction followed by the *to*-infinitive indicates a future state of affairs (lines 06–07). In line 09, J offers DA the opportunity to argue his position, and he brings up not only prior opinion but also *assuming facts not in evidence*, the objection that stimulated J to remove the jury in the first place. That is, in the adversarial system only sworn testimony from witnesses in court counts as evidence, not testimony from future witnesses (witnesses who have not testified such as detective Rigolo) or questions from attorneys. He concludes with specific questions PA can ask WS (lines 15–17). In line 20, J gives PA a second chance to argue her position and once again she uses the identical *be going* + *to-* infinitive as a general statement about impeaching testimony (lines 22–24).

In line 25, J begins her response to PA immediately with no pause, latching onto the end of PA's basis from line 24. Turn initial, contrastive *Well* indicates a judicial challenge to the basis (while simultaneously dismissing it) and, in contrast to PA's general claim, J rules specifically (or question by question). In this regard, J proceeds to quote PA's last question from cross-examination (lines 25–27). At quote onset J looks down at what appears to be a written text of the question about detective Rigolo. In so doing, she fosters the impression that the improper question was important enough to write down. At the end of the quote (over distraught mood) she gazes up from the text and towards PA to state: that is clearly an improper question. J not only endows her words with more epistemological significance and authoritative power through the direct quote. The epistemic stance adverb clearly expresses absolute certainty about the proposition, conveying an "obvious" meaning with some degree of residual evidential effect (since the impropriety is easily verifiable).

But looking at J's words alone misses much of the meaning she conveys. She mobilizes improvisational co-speech beats in a rhythmic drive to enliven her words and coordinate the dynamic unfolding of interaction. Like an orchestra conductor's baton, beats move up and down to parse speech into prominent visual segments. They orchestrate the rhythmic pulse of speech and foreground key segments of information, increasing the oratorical power of the message. According to Krahmer and Swerts (2007: 396) speech synchronized beats increase the persuasivenss of the message compared to words, phrases or clauses without them. Maricchiolo et al (2009: 244) found similar results: "Experimental design with control groups shows that beat gestures have an important effects on receivers' perceptions of communicative effectiveness, and persuasiveness of the message."

With these points in mind, J beats out the rhythm of her response to PA (lines 25–27) in a "precision ring" gesture, where the tip of the index finger touches the tip of the thumb to form a ring shape (with the other three fingers spread out, see Figure 2.5 below). According to Muller (2014:118) this gesture "makes a specific reference to something . . . in contrast to something more general," and Morris (1977: 58) finds that the speaker uses such a gesture "to express himself delicately and with great exactness." Additionally, her precision beat gestures exhibit marked variation based on distinct contextual coordinates to make the quote more emphatic sounding: short oblique angles prior to and after the direct quote, and much steeper (vertical up-down) angle with higher elevation on the upstroke and marked jerk on the downswing over the direct quote. After the quote, shifts from vertical to horizontal precision beats (two lateral gestures with palm down in line 27) further intensify J's negative position on the *improper question* and indicate that nothing further needs to be said (Kendon 2004: 256–259; Calbris 2016). In sum, the direct quote adds epistemological significance and authoritative power to her words, especially when fostering the impression of reading from a written text. The precision ring makes a sharp effective point, provides a precise representation of PA's words. Beat gestures highlight main points of the argument and accentuate rhythmic affect. And the stance marker, *clearly,* adds certainty to the proposition that the question is obviously improper. Together, all four semiotic resources mutually elaborate one another to add considerable authorial power to her utterance.

Figure 2.5: lines 25–27 *precision-ring gesture.*

The precision beats illustrate our prior points about speech-gesture synchronization and how gestures capture meanings not represented as economically in speech. First, the 0.3 second pause and gesture hold co-occurring with *clearly* demonstrate in vivid detail how J maintains the integrity and close temporal coordination of the speech gesture ensemble. Gesture stops when speech stops, and then starts back up when speech starts up. And second, the image of precision and no further discussion are not available from J's words; only her gestures convey such representations.

And there is another aspect to the precision ring gesture. Both Lempert (2017) and Muller (2014) emphasize it possesses contrastive focus. For Muller (2014) J's precision gesture delivers a precise argument about how she rules on evidence, one that contrasts with PA's general argument about impeachment: precise gestures for a precise point about a precise question. For Lempert (2011: 241; 2017), precision-grip gestures make "a sharp effective point" and simultaneously index the speaker as "being argumentatively sharp. In Lempert's analsysis however the nonselected fingers are curled toward the palm rather than spread out as in Figure 2.5. Both authors maintain that such beats are metapragmatic, making a sharp effective point and/or precise argument that shapes interpretation of itself.

Even so, J's oratorical gem makes little impact on PA, who asks "why" her question is improper (in line 29): *Could the court explain to me why?* Appearing taken aback by the question (notice the delay and error-correction), J's response produces a further lamination of participation when she asks the court reporter to read back DA's basis for the objection in lines 31–32 (now J and CR, excluding other participants from participation). At a finer level of granularity, CR is typing on the steno machine, recording the proceedings, when J asks him to read back DA's basis for the objection. CR stops recording and scrolls up the steno paper (with two hands, see Figure 2.6 below) in search of the basis for over 15 seconds, then stops and looks at J to ask: *the basis for the objection?* (line 35). Instead of answering CR's question, J repeats PA's question on detective Rigolo. But as she begins her utterance, *The question that Miss Lasch-*, CR raises his finger for her to stop till he can scroll back down the steno paper in position to transcribe. Notice that J cuts-off the address term (*Miss Lasch-*) and waits an additional 4.5 seconds till CR is ready. In essence, J attempts to close down this new lamination of participation but CR opens it back up by raising his finger for J to stop while he readies the steno machine. This demonstrates in vivid detail the dynamic interplay between speech, gesture and material objects and how machine participation figures prominently in legal interaction.

Figure 2.6: line 33 *CR scrolling up steno paper.*

On lines 41–43, J repeats the quote from lines 25–27 on the evidence in question. This time, however, rather than evaluate the improper question or provide the basis she turns the tables on PA by posing a question of her own (in line 45): *Do you agree with mister Black that* detective Rigolo has not testified. Although J uses a yes/no question, PA fails to reply with the expected yes/no answer, indicating instead that detective Rigolo will be testifying on rebuttal. As a result, J interrupts PA's answer prior to completion (over the adverbial) when she recognizes the expected *yes/no* answer is not forthcoming, and then repeats the question in overlap (notice also J's canonical self-initiated, self repair (with cut-off) performed verbally and visually with a vigorous headshake and closed eyes). After turn initial *No*, PA attempts to expand her answer, but J interrupts again with another question (line 55), an emphatic question consisting of increased volume, stress and lateral head shakes: *Then how in heaven's name is that a proper question.* The lateral headshakes over the *Heaven's name* idiom make her words even more forceful.

In a dialogic lamination of participation, J's questions and interruptions teach PA how to cross-examine a witness in the here and now as she questions her about the represented event, a *yes/no* question that requires a simple *yes/no* answer. Indeed, the only control and asymmetrical relationship that emerges is not between attorney and witness in examination but between J and PA, where the former imparts a stern lesson about control in conference. In more theoretical terms, Silverstein (1998: 266) refers to this process as a symbolic interaction in which "what was said maps onto what is done" or a "mapping of denotational text to interacting text." In a similar vein, Wortham (2001: 138) shows how "narrator and interlocuters act out relationships partly analogous to those represented in the narrative." In the case here, we see another intricate lamination of participation in J's cross examination of PA, consisting of a mapping of denotational text to interactional function or what we are referring to about the past becomes what we are doing in the here-and-now – how the represented event is positioned or enacted in the current speech event. By interrupting and repeating the question, J demonstrates the proper evidential procedure for cross-examining a witness, how to build a leading question instead of giving an argumentative speech. J not only instructs PA how to question a witness but how to answer (and obtain an answer). Indeed, J and PA simultaneously mobilize the problematic objection event not only in historical representation but also in current performance.

After PA responds to the *heaven's name* question (line 58) it becomes clear that it was not the reply J was looking for. In fact, J was not looking for a response at all. For her emphatic question was actually, it turns out, a rhetorical question not designed for a response. In line 62, J addresses PA by name and, first, delivers a noticeably marked pause of close to 11 seconds (line 62) consisting of several lateral head shakes with closed eyes in an intensification of disbelief that PA still cannot figure out the improper violation and continues to resist J's ruling (see Figure 2.7 below); and, second, she escalates the "cross-examination" with an insulting reprimand that undermines PA's competence as an evidence expert: *this is not a course in trial practice.* Several lines later (lines 66–67) she delivers a similar insult: *Now I'm not going to go through a course.* Indeed, only a novice law student – not a formally trained attorney – would fail to grasp the transparent simplicity of such inadmissible evidence and improper questioning. Put more prosaically, J looks like one of Mertz' (2007) law school professors imparting a stern lesson to a recalcitrant student through the Socratic method. Indeed, the institutional reflexivity of legal discourse works its way through the objection conference to make explicit attacks on credibility and professional identity rather than solely emphasize rules of evidence (as in the objection sequences from the prior section), doubtless why J dismissed the jury in the first place.

Figure 2.7: line 62 *closed eyes in lateral head shake.*

In lines 67–70, J continues the narrative that questions PA's limited knowledge of evidence law and legal competence but allows space for the possibility – through the *either/or* correlative conjunction – that her question may have been a tactical maneuver. Even so, J recruits closed eyes and lateral head shakes positioned at an oblique angle over both alternatives, a more emphatic index of not just negation but disbelief as well (see Figure 2.8 below). As we saw previously, the head tilt indicates a symbolic intensification of negation, something amiss or imbalanced (Calbris 2016: 96–98). And here J indexes PA's blatant disregard not only for evidential propriety in the first place but for all subsequent attempts to correct her violations. She never figures out the problem, and J treats it as the "last straw" in a litany of attempts to reason with her.

After that, J ends the conference with specific commands to PA. She contextualizes her verbal and visual crescendo with the emphatic discourse marker *now* (line 70) that transitions into a new topic, from the general admonishment to specific

Figure 2.8: line 68 *head tilt.*

directives. And the directives unfold in and through a polyrthymic metricalization of interwoven themes.

We saw previously how J used precision beats to accent speech rhythms and articulate metaphoric representations. In lines 70–74, she uses inter and intra digital beats – a dialect of the fingers – to add affective intensity to and count off her verbal directives in a rhythmic resolution of the objection conference. Interdigital beats occur when the tip of the index finger on the right hand lands on the tips of the little finger, ring finger and middle finger (respectively) of the left hand in an ascending progression of interdigital increments (see Figure 2.9 below). Each subsequent digit unfurls incrementally prior to being deployed. Intradigital beats occur with the index finger of the right hand latched onto the middle finger of the

Figure 2.9: line 71 *interdigital beat gesture on little finger.*

left hand, beating out rhythmic significance over an entire idea unit in this latching position (line 74). She uses both types of beat gestures to coax sharply articulate rhythmic commands out of her two hands and fingers.

In line 71, interdigital beats co-occur with verbal repetition consisting of the future modal and negative particle followed by the phrasal modal (*will not be asked*), a parallel structure that unfolds as follows: first, a short upstroke of the index finger on *his*; second, the index finger hits the little finger on the first *opinion testimony*; and third the index finger pushes down the little finger on *other*, with a .6 second hold till *experts* arrives. After a 2.9 second pause in line 72, the index finger upstroke on *He* maintains a high elevation hold for 3 seconds (see Figure 2.10 below), then after a steep (seventy degree) angle dive, hits the ring finger on *facts* and pushes down on *evidence*. After a lengthy post stroke hold (typically used for emphasis), the final interdigital beat (line 74) lands on *free* before it evolves into a series of intradigital beats. By individuating each item in the list through gesture, J cumulates and expands the directives and shows that each item is important enough to draw attention to.

Figure 2.10: line 72 *high elevation upstroke on interdigital beat to ring finger.*

Her beats participate in a text metricalized structure of speech and gesture that foregrounds the opposition between the two specific negative directives and a general positive directive, a perceptually salient poetic performance that not only calls attention to itself but also guides evaluation of the unfolding text (lines 70–74). In the case here, J's multidimensional pattern of consistency and variation in gesture and speech calibrates text interpretation and signals that the otherwise disparate elements should be compared and linked as a unit. And notice an embedded polyrhythmic lamination within the parallel structure: staccato rhythm in ***any*** *other witness* co-occurring with staccato beats over each lexical item. Precise synchronization in such multidimensional rhythms adds further emphasis to her directives as it builds to a dynamic crescendo.

Why does J use interdigital beats in lines 70 and 72 before finishing the parallel structure with intradigital beats in line 74? By distinguishing each specific digit, she makes the interdigital beats correspond to specific rules of evidence (to opinion testimony and facts not in evidence), using multiple digits for unpacking specific details. On the other hand, intradigital beats mobilize a more general structure for a more general idea or directive, using a single digit for a single idea unit. Together, these multi-textured cross-rhythmic gestures work in concert with speech to maintain emotional depth and evidential clarity to ensure PA understands (and complies with) the directives.

2.4 Summary

This chapter offers insight into the organization of a novel form of institutional questioning practice and the role of multimodal conduct in shaping that organization. In over ten minutes of trial interaction we have seen neither the putative prevalence of question-answer pairs nor much in the way of witness control. Instead, we have seen several minutes of objection-mediated forms of participation that evolve into several minutes of conference. In the first part of the chapter, the witness varied his involvement in the answer role and recognition displays through embodied and paralinguistic forms of residual participation, even though the witness role was officially excluded in the objection participation frame. In the second part, we did see questions, answers, and institutional control but not between attorney and witness. By using cross-modal poetic forms J provides the only effective cross-examination in both segments and constructs professional identity in a highly animated affective display, a multimodal mapping of denotational text to interactional function.

By the same token, that construction relates to participation roles governing not just questions, but proper questions, shaping their interpretation and trajectory. Most importantly, participation roles take precedent contingently over sequential patterning, filtering questions and bleaching answers to meets strict canons of evidential relevance. Objections problematize questioning practices while the objection conference not only imparts rules dealing with improper questioning but also evokes institutional reflexivity to shape evidential credibility and professional identity. Put simply and stipulatively, what Silverstein (2014) refers to as *denotationally explicit metapragmatics* plays a key role in legal discourse – foregrounding the nature of proper questioning and its relation to inappropriate conduct – and points to ways we might thinking about communicative practices more generally (thing about complaining about complaining or complaining about other speech acts or complaining about someone being a "busybody" and so on), keeping in mind of course that a key institutional signature of objections is that the ultimate evidential authority rests with the judge. They reveal how participants *do things* with felicity conditions as a contingent and practical evidential accomplishment.

Let us return briefly to the concept of speech exchange system – differential participation rights based on institutional role discussed in chapter 1. Indeed, looking at speech alone appears to reveal, *prima facie,* the asymmetrical distribution of discursive options as orthodoxly conceived, but looked at multimodally a quite different vision emerges. The concept ignores the improvisational complexity of courtroom interaction, as the defendant takes a multimodal stance on the attorney's questions, something prohibited in speech.

Our goal however is not to deny the pervasive patterning of question-answer pairs in courtroom examination. Nor does the analysis here diminish the import of institutional control that emerges from questioning practices. Rather, we hope to build objections into the very fabric of those patterns, providing a deeper, multimodal analysis of those patterns, instead of simply relegating objections to deviant status or treating them as tangential anomalies. Most important, the data here reveals the necessity of incorporating multimodal conduct such as gesture into any comprehensive analysis of trial interaction. Indeed, it is hard to fathom how we could understand much less analyze empirically legal ritual any other way after looking at the massive contribution of multimodal conduct in objection mediated practices.

In the next chapter, we continue our investigation of multimodal oratory by looking at another neglected area of legal ritual: How opening statement may be transformed into an evaluative argument in a stunning departure from its orthodox role as an outline of forthcoming evidence.

3 *She does not flee the house:* Poetic rhythms of space, path, and motion

Unlike closing argument of a trial, opening statement is not an argument that comments on evidence and the credibility of witnesses. Instead, opening merely outlines the trial attorney's case for the jury, providing context for upcoming witnesses, evidence and testimony (Heffer 2005: 75). Even so, the line between argument and outline is a blurry one and often the interpretation of the boundary rests with the judge in the case. As Perrin et al (2003:121) put it, the prohibition on the argument rule is "ill-defined and poorly understood. The actual application of the rule... varies from jurisdiction to jurisdiction and even from judge to judge." That discretion often rests with the judge figures prominently in case strategy, especially given the importance of opening for trial outcome.

According to the literature on trial practice and forensic linguistics, opening statement represents the most important stage of the trial (Cotterill 2003: 65). It frames forthcoming evidence in the case, shapes the first impression of jurors, and plays a significant role in their final decision. That is, a favorable initial impression makes a lasting impact on jurors; favorable impressions formed in opening statement correlate with a favorable vote in the jury's ultimate decision in the case (Mauet 2017; Haydock and Sonsteng 1990).

And those impressions are managed in and through the attorney's skilled use of language. Cotterill (2003) examined "strategic lexicalization" during opening statements in the OJ Simpson murder trial, and how lexical choices differed between prosecution and defense. On the one hand, the prosecuting attorney referred to Simpson as a "controlling and abusive husband" engaged in a progression of domestic violence against his spouse that escalated to her ultimate murder. On the other hand, the defense attorney claimed that Simpson's actions were isolated and unrelated "incidents", domestic altercations typical of most married couples. Such competing descriptions of the same behavior functioned as interpretive frames for shaping the jury's perception of the case. In a similar vein, Chaemsaithong (2018, 2017a,b) discovered that, even though they cannot comment on the credibility of evidence or witnesses during opening, attorneys "bypass" such constraints through footing, reported speech, and evaluative stance. In the example below, he (Chaemsaithong 2018: 4) demonstrates how evaluation of witness credibility materializes in the defense attorney's direct quote, a quote delivered in a contrastive format without any "affective intonation" (a surprisingly stoic representation of the unsettling sweep of events that just transpired at the defendant's home): "She [the victim] goes into the house... and makes a call to her friend Ann Mercer, who is an acquaintance. That's the first time they have ever gone out together... She doesn't call anyone in her family... the police... any relative but she calls Ann Mercer, and says, "I've been raped. Come and pick me up." (See Matoesian 2001: 113). Both studies demonstrate how attorneys "make the opening statement essentially argumentative" through their persuasive manipulation of language.

But there is much more to opening statement than language. In this chapter, we analyze how opening statement manages to "bypass" argumentative constraints through a dynamic integration of speech, gesture, and material objects, a multimodal configuration that encodes motion events crucial to the case. We demonstrate how the defense attorney organizes opening statement through an iconic gesture that traces the witness's movement along the surface of a crucial piece of evidence – a diagram of locations in the suspect's residence. In so doing, we show how a multimodal ensemble of rhythmic resources consisting of speech, exhibit and gesture shapes spatial images into relevant objects of evidentiary knowledge, into inconsistencies in the "anticipated" evidence and an argument for a favorable verdict for the defense.

3.1 Multimodal conduct in language and law

Trial practice textbooks typically devote considerable space to the use of gesture, exhibits, and other forms of embodied conduct as necessary features of persuasive delivery in general and opening statement in particular, a consideration that may be instructive for linguists studying trial discourse and other legal-institutional practices. According to Mauet (2017: 98), effective opening statement should use "body language, hand and upper body gestures to provide emphasis and interest" in addition to paralinguistic features such as "varying volume, pace, pauses and silence" (2017: 95. In a similar vein, Haydock and Sonsteng (1990: 326) recommend that attorneys not only use "effective gestures" in concert with speech but also employ "visual aids and exhibits that significantly increase the effectiveness of opening statement" (Haydock and Sonsteng 1990: 312). Mauet (2017: 95) finds that exhibits and other visual resources increase the "impact and retention" of information, "especially at the beginning of the trial." Indeed, not only do exhibits make information "easier for the jury to understand (Haydock and Sonsteng 1990: 312), "they also tell jurors right from the beginning of the trial that you care about the case and care that the jurors fully understand what happened and why" (Mauet 2017: 342). In the trial here only the defense used an exhibit during opening statement. Cooperating as a multimodal ensemble, "all three components – verbal content, nonverbal delivery, and body language – must work together before an opening statement will be persuasive" (Mauet 2017: 99).

As noted in the introduction, multimodal conduct includes not only the interplay of speech, gesture and gaze but also material objects in the production of emergent forms of socio-legal meaning. From this perspective, materiality is embedded in and relevant to the construction of social action. According to Heath and his colleagues (Heath, Hindmarsh, and Luff 2010) material resources typically occupy a passive role in orthodox social science research – as physical props of the social setting or environmental conditions that encapsulate social action. But such an orientation ignores how participants mobilize material objects to index relevant features of context at specific moments within unfolding courses of action. Rather than view materiality as a passive reflex of physical environment, we follow Heath et al to demonstrate the active role of material conduct in the construction and co-construction of meaning in opening statement, in particular, how material conduct figures prominently in the performance of legal action and interaction.

The chapter is organized as follows. In the next section we provide a brief overview of how an attorney introduces and choreographs an important exhibit, shaping it into a focus of joint attention and object of evidentiary knowledge. After that we demonstrate how the encoding of motion events functions more efficiently and effectively – especially in the alteration of route paths – through the integration of speech, tracing gestures, and exhibit than through speech alone. Each modality mutually elaborates the other in the encoding of motion events and path trajectories to anchor inconsistencies not only in forthcoming testimony but, just as importantly, in the here and now opening statement. In so doing, the attorney transforms diffuse territories on the diagram into a sharply focused and concentrated field of motion evidence – into an argument about witness credibility.

3.2 Introducing and contextualizing the exhibit

According to defense attorney Roy Black, that the victim kissed the defendant and removed her pantyhose in the car was proof of foreplay and that her actions after the alleged rape, especially that she stayed at the Kennedy estate rather than flee, provided further evidence of consent (or at least a striking anomaly of post-rape behavior).

In her opening statement, prosecuting attorney Moira Lasch stated that Bowman went into the Kennedy estate for a tour of the home. Defense attorney Roy Black, in response, states: "The prosecutor would have you believe the complainant went into the house for a tour of the estate as a tourist would do. That's simply not what happened. When they're in the car there's kissing going on. The complainant winds up taking off her pantyhose." Black then introduces a diagram – an iconic representation – of locations in the home.

Example 1

001	Perhaps at this time a- we have the diagram
002	(.1) o::f the home
003	that perhaps it would be easier for you to understand
004	the evidence if I could use it
005	((36.0 break while attorney and court clerk position easel and diagram))
006	this is (.3)
      [
      ((manual positioning of easel toward viewing field of jury))
007	>>move this out of the way<<
      [
      ((manual movement of podium out of the viewing field of jury))
008	 (.5)
      [
      ((walks toward diagram, gaze toward diagram))
009	a diagram (.)
      [
      ((right palm fingers spread across front image on diagram))
010	of what the home looks like
      [
      ((axial rotation of body toward jury, gaze shift toward jury while
      standing in front of diagram))
                ((break of several lines/22 seconds))
011	During the course of the case
      [
      ((right hand lifts off diagram fingers spread))
012	uh we will show you uh
      [
      ((right hand grabs pointer, left hand unfurls toward jury))
013	in fact I'll show you
           [
           ((left hand moves toward chest))
014	now some of the photographs
      [
      ((left arm extends toward jury, palm supine))
015	that we've taken and one or two uh
016	in order to give you
            [
            ((beat))
            ((head beat))
017	a *real* sense:: (.)
      [[
      ((oblique beat))   ((oblique beat))
      ((head beat))   ((head beat))

018   of what   the home was like
       [[
       ((micro oblique beat))   ((micro oblique beat))
019   and what the environment is like there
                  [
                  ((oblique lateral toggle movement))
020   because it's   so   important
              [[[
              ((oblique beat)) ((oblique beat)) ((oblique beat))
021   to this case
       [
       ((L and R hands come together holding pointer, body torque and gaze toward diagram))
022   (.) to first of all
023   orient yourself as to
024   where everything is
025   cause its   very hard to understand the testimony
       [[
       ((micro lateral head movement))   ((palm beat above diagram))
026   until-you   unless you   get   a   good
       [[[[[
       ((beat))   ((beat))   ((beat)) ((beat)) ((beat))
027   idea   that's   also   important
       [[[[
       ((beat)) ((beat))   ((beat)) ((beat))
028   to   physically   see
       [[[
       ((micro beat)) ((micro beat)) ((micro beat))
029   u::h ah
030   parts   of   the home ((entire line staccato))
       [[[
       ((beat))   ((beat))   ((beat)) ((all three beats two-hand staccato))
                                          ((all three beats high elevation upstroke))
                                          ((all three beats low bottom downstroke))
031   (.) because   they're   really
       [[[
       ((upswing of gesture)) ((downswing beat))   [((beat))
032   critical (.) to the understanding of this case
       [
       ((beat))

In ideal-legal terms, a diagram is not "real" but "demonstrative" evidence (in more prosaic terms it is not the "real" thing but represents the "real" thing). That is, it does not refer to an actual event or piece of evidence like a gun or contract but, instead, "is used in conjunction with witness testimony to clarify cumbersome, complicated, or confusing testimony" (Perrin et al 2003: 252–253). According to Mauet (2017) diagrams must be introduced and "choreographed" for the jury with a "dramatic flair." While Black argues against the prosecution's "tour" explanation, he proceeds to use the diagram for his own "tour" of the home (at eleven minutes into opening), using it to impeach Bowman's credibility under the auspices of giving the jury a "neutral" inspection of locations in the home. In line 001, he starts by introducing a diagram of the home, and then instructs the court clerk and co-defense attorney to position the easel and diagram for the jury. But why go through the time and effort to move the easel and diagram – two big objects – into place? (See Figures 3.1 and 3.2 below) If the exhibit is of such importance why not have it already assembled and positioned? Rather than treat the material object as a passive reflex of physical context, Black's orchestration of movement and speech locates the diagram as an agentive participant – one that can be felt, seen, and experienced. For the jury, such material conduct must be important enough to interrupt the flow of opening statement. Combined with deictic "this" in line 006 and positioning the diagram towards the perceptual field of the jury, he foregrounds and draws attention to it. While maintaining manual contact with the diagram, Black contextualizes the importance of the exhibit so jurors can see "what the home looks like" (line 010). Just as deferentially (and spontaneously), the stance adverb *perhaps* in lines 001 and 003 offers the diagram to and for the jury – for their benefit.

Figure 3.1: (line 005. Defense attorney Black is on the left).

On closer inspection, notice how Black contextualizes "importance" through a medley of stressed stance markers and intensifiers in lines 017, 020, 025, and 031–032 (*real sense of what the home was like, so important to this case, very hard to understand the testimony, physically see parts of the home because they're really critical*), all designed to give jurors a *physical sense* and *idea* of the home because it is *critical to the under-*

Figure 3.2: (line 009 *a diagram*).

standing of this case (line 032). Just as important, he foregrounds the main points in his statement by integrating speech with beat gestures (in lines 016, 017, 018, 020, 025, 026, 027–032) to not only orchestrate the rhythmic pulse of speech but also emphasize the main pieces of information – increasing the persuasiveness of the message in the process. As a poignant illustration, in line 30 Black delivers each beat with two hands – high elevation on the upstroke and low bottom on the downstroke – and staccato like rhythm synchronized with his staccato speech segments to further emphasize and foreground key points in the forthcoming narrative (*parts of the home*) (See Figures 3.3 and 3.4 below).

Figure 3.3: (line 030 beat gesture upstroke).

Figure 3.4: (line 030 beat gesture downstroke).

Black's improvisational choreography shows how material objects are relevant at specific moments in the unfolding interaction to accomplish distinct evidential tasks in opening statement. Rather than use the diagram to merely help the jury *understand the evidence* (lines 003–004), his contextualization work invokes locations and movements in the diagram to *construct* inconsistencies in Bowman's testimony. As we will see in the next sections, Black's descriptive "tour" of the home reveals how speech, gesture and the diagram mutually and reflexively constitute one another, anchoring crucial points of evidence and witness credibility.

3.3 A multimodal poetics of space, path, and motion in three parts

3.3.1 Staying in the house: Departure from normative expectations (lines 001–005)

According to Bowman, after the sexual assault she went back into the house to call a friend to pick her up. In the meantime, the defendant came back into the house to have a "discussion" with her about the incident. During opening statement, defense attorney Roy Black mentions that rather than *flee* the house after the assault she stays in the home. Black's narrative picks up on the theme at 20:14 minutes into his opening statement.

Example 2 (D=Diagram, J=Jury, RH=Right Hand)

```
001    After thi:::s (0.6) discussion (0.5)
       [
       ((gazes at D and points to room in diagram (D) with right index finger on *this*,
       and shifts gaze back to jury on vowel stretch and 0.6 pause))
002    she does not flee the house (.)
              [
              ((flinging point beat w/right hand to diagram))
              ((lateral head shake))
              ((each gesture stroke in lines 2–5 lands on the main verb))
003    she does not go out into the parking lot (.)
              [
              ((gaze to D with flinging point beat w/right hand to D))
004    she does not get into her        car and leave (1.1)
              [                         [
              ((same beat w/RH to D))   ((gaze returns to jury))
005    She **sta::ys** (0.4) in the house (1.4)
              [
              ((noticeably marked pointing beat w/RH to D while still gazing at jury))
              ((marked gesture recoil and head jerk on *sta::ys*))
006    As far as we can determine (.)
              [
              ((gaze to D, index finger point with RH at room in D with 1.5 second
              post stroke hold)),
007    she not only (.) (wi)- in this room
                           [
                           ((three beats with right index finger))
                           ((single parenthesis above is transcription doubt))
008    she describes (.) where she's in
       [                  [
       ((gaze back to J)) ((2 right index finger beats))
009    and has this conversation is
                   [
                   ((1 right index finger beat))
010    right over here (.)
       [
       ((gaze to D))
011    Right next to the back door (.)
                      [
                      ((RH index finger point to D))
                      ((traces finger to back door and back to room))
012    She not only does not go out the back door (0.3)
       [                              [
       ((gaze back to J))             ((finger traces to back door))
                          [                          ]
                          ((lateral head shake))
```

```
013    she goes         through the kitchen
              [              [
              ((gaze to D))   ((tracing gesture on surface of D down several inches))
014    through the dining room
              [
              ((tracing gesture with index finger on surface of D to right several inches))
015    **into the living room**  (1.1)
              [
              ((tracing finger with index finger moves slightly off D to right and lands on the
              living room location))
              ((gaze returns to J on "room"))
              ((finger stays on "living room" location))
016    where she (.) rummages around (.) the property in there (.)
                   [                        [
                   ((index finger beat))    ((index finger beat))
                        ((with finger still on surface of D))
017    and picks up a photograph (.) that she says she's taking with her
              [                              [
              ((right hand open palm up))    ((open palm decays in upward movement))
((using the moving figure in the house as the deictic center in lines 013, 014 and 015, Black's finger
traces (1) down (2) right and (3) further right.))
```

In lines 001–002, Black refers to a problematic feature of the prosecution's case: why, after the rape incident, did Bowman go into and stay in the house rather than flee? He begins by gazing at and gesturing toward the diagram of the house, turning it into a relevant object of evidentiary knowledge – of joint attention. Notice at the outset the interaction among gaze shifts, pausing and vowel lengthening mobilized to distribute distinct forms of participation and involvement simultaneously. Although Black starts line 001 by gazing at and gesturing toward the diagram, he shifts gaze (to the jury) during vowel lengthening and a 0.6 second pause to fully engage jurors prior to making the main points in the ensuing argument.

And he delivers those points in a multidimensional poetic performance, a perceptually salient form of text-metricality that not only calls attention to itself but also calibrates text interpretation and signals that otherwise disparate segments should be compared and linked as a unit. As mentioned in the introduction, in Jakobson's (1960) classic work on poetics, repetition with variation produces a metapragmatic affect that enlives and emphasizes the message. Poetic patterning consists of measured repetition, lexical recurrence, and contrastive opposition in the aesthetic text, rhythmic density that adds emotional intensity and normative depth to his evaluative frame. As we will see, such similarity driven and perceptually salient patterns of recurrence apply to co-speech gestures as well as speech. In terms of bypassing constraints on argument, metricalized speech and gesture create a sense of cohesion in the message and naturalize normative expectations (making it appear normal that women should flee rather than stay) in the negotiation of an agentive identity.

Black's dense weave of poetic patterning begins on line 002 with the third person, *do*-auxiliary plus uncontracted negative, manner verb (*flee*) and path complement (*the house*). Line 003 repeats the third person, *do*-Aux, and uncontracted negative while it adds the directed motion verb (*go*), spatial particle (*out*), and directional path that encodes a boundary crossing event (*into the parking lot*). And in line 004, he repeats the third person, *do*-Aux, uncontracted negative, and directional path (*into her car*), adding the *get* plus *leave*-motion verbs. This follows Talmy's (1985) influential work on the encoding of motion events in language, more specifically, that motion events consist of four components: a moving object or figure; the type and manner of motion; the ground through which the figure moves (from a source or origin to a goal or destination); and path or direction of the figure's motion.

At a finer layer of detail, several observations are relevant to framing normative expectations and witness credibility in Black's encoding of motion events. First, the uncontracted negative (*does not*) conveys semantically focal and evaluative information and, unlike the contracted version, emphasizes negation (and disagreement), especially in adversarial settings. Second, although *flee* in line 002 indeed encodes manner of motion it does quite more than that. That is, it not only encodes motion out of an enclosed setting, conflating manner and path of motion, but also denotes, according to Huber [(2017: 43)] "speed, fear, or danger": running away from some threat. Third, the subdivision of listing items (lines 002–004) recalibrates the looming opposition (in line 005) into a more striking – numerically amplified – violation of normative expectations (rather than just saying, for example, "she doesn't leave"). That is to say, spatio-temporal progression (lines 002–004) contributes to the natural "feel" of cohesion that contextualizes contrastive opposition, an opposition that foregrounds normative preferences and Bowman's departure from those expectations.

More explicitly, in such circumstances one should first *flee*, second *go out into the parking lot*, and last *get into her car and leave*. Of course, her "departures" as framed by Black are marked. As Givon (1978: 109; see also Horn 1979) puts it: "Negatives are consistently more marked in terms of discourse-pragmatic presupposition, as compared to affirmatives . . . negatives are uttered in a context . . . where the speaker assumes the hearer's belief in – and thus familiarity with – the corresponding affirmative." And, in lines 002–003, Black contributes to such feeling using a discretely bounded tempo with each line punctuated by noticeable silence, while mobilizing the long pause of 1.1 seconds before the second half of the contrast in line 005.

The actual behavior that departs from those expectations emerges in the second half of the contrast in line 005. After the lengthy pause of 1.1 seconds and gaze shift back to the jury, the contrastive opposition reaches a marked resolution with *She sta::ys (0.4) in the house (1.4)*. Notice how the durational (or exist) verb, (Levin 1993) with a non motion meaning) verb – encoding a continuing state of stationary location – co-occurs with vowel elongation, marked stress, and lengthy pauses (0.4 and 1.4 seconds) as an icon of staying in the house for a period of time (or it could be a time passing metaphor with heavy iconic loading). In so doing, Black recruits a glaring rhythmic contrast to foreground and evaluate Bowman's violation of the normatively gendered order: after being raped, a woman should flee.

However, looking at speech in isolation from other semiotic modalities leaves us with an incomplete view of evaluation work in opening statement. In line 002, Black synchronizes the motion verb with a lateral headshake, signaling a sense of disbelief or confusion over Bowman's behavior. Simultaneously, he begins a series of flinging path gestures toward the diagram with the extended right arm and index finger, beating out the rhythm of his pointing gesture (a rhythmic gesture that accompanies the deictic gesture) from lines 002–005 (See Figures 3.5 and 3.6 below). Each gesture stroke lands on the main verb in lines 002–005. In lines 003 and 004, he points toward the parking lot in the diagram (once again, on *this* in line 001 he points to the room in the diagram where the conversation occurred. In so doing, Black mobilizes not only lexical recurrence but an aesthetic interplay between speech and gesture that adds a feeling of polyrhythmic agility to his performance.

Figure 3.5: (line 003 *go out*).

We mentioned previously that the manual modality encodes aspects of the message visually or imagistically, additional images often unavailable in speech alone. We also mentioned above that verbs like *stay* do not encode movement (that is, it pos-

Figure 3.6: (line 005 *stays*).

sesses a non motion meaning) but denote a stationary locative event (which is why Jackendoff (2002) refers to it as a "maintaining verb" and Levin (1993) an "exist" verb). However, the off-kilter flourish of sound, gesture, and motion in lines 002–005 preserves gestural continuity by repeating the same gesture over the main verbs in each line, recalibrating *stay* as agentive motion in the process (she has a choice of staying opposed to actively fleeing). Black achieves such continuity by maintaining the integrity of the parallel structure and repetitive gestures, equating *stay* and the gesture of actively *staying* on the same motion plane as *flee, go out, get into* and *leave*. Each gesture in lines 002–005 co-occurs with the main verb and consists of a flinging-path beat and rhythmic point (with extended index finger) toward the diagram. Each gesture encodes the figure's normatively expected motion along a path trajectory from a source to a goal like *flee the house, go out into the parking lot, get into her car/leave* and *stays in the house*. Indeed, its participation in a sequence of repeated gestures recalibrates *stay* as a motion verb in contrast to its stationary or existence status depicted in speech alone. (A marked gesture recoil and head jerk accompanies *stay* to further emphasize its motion meaning status rather than static locative relationship). Looked at in speech alone, *stay* appears as a stationary locative possessing a non-motion meaning. Looked at multimodally as part of an ensemble of similar gestures, it is not a static but dynamic motion verb with agentive meaning. Thus co-speech gestures demonstrate how complex rhythmic structures are folded into interwoven themes that recalibrate the meaning of dynamic motion events, text-metricalized meanings depicted inadequately through analysis of speech alone. If this is so, then empirical examination of speech-synchronized gestures can lend analytic insight to orthodox linguistic interests like the framing of motion events.

3.3.2 The back door: A crucial parenthetical (lines 006–011)

In line 006, Black produces another striking integration of modal resources. *As far as we can determine* (see Figure 3.7 below) encodes a less than certain stance about Bowman's location and movement in the home. However, at utterance onset, he points (with index finger extended with other fingers retracted under the thumb) to the location in the diagram while simultaneously gazing at the target of the point. The index finger extension point (including the post-stroke hold) and gaze not only provide definite reference or locate objects in space but also function as evidential resources to impart more authority and certainty than his words indicate. While his verbal component is less than certain, his gestural deictic and gaze more than compensate by reifying the diagram as the source of information, disambiguating and upgrading referential certainty. The first person plural (*we*) adds a further degree of disambiguation through epistemic corroboration in pronominal incorporation, producing a state of joint visual attention and instructing the jury to collaborate in object location. Rather than merely assuming the classic distinction between the grammatical encoding of epistemic (degree of certainty of the proposition) and evidential (information source) functions (as per Aikhenvald 2004: 3–4), we show their close relationship when looked at multimodally (see Roseano, Gonzalez, Borras-Comes & Prieto 2016).

Figure 3.7: (line 006 *as far as we can determine*).

Next (line 007), Black projects a correlative conjunction to indicate Bowman's failure to leave the house. However, after completion of the first part he aborts the second with a cut-off "hitch" that alters utterance trajectory. As it turns out, the altered trajectory takes the form of a clausal parenthetical about Bowman's location relative to the back door exit of the house. The parenthetical interrupts the flow of Black's utterance in progress to, first, grab the jury's attention and, second, add crucial commentary to the sense and significance of his communicative strategy. Just as interesting, the parenthetical not only delays utterance progressivity but also halts his gesture in progress, a tracing movement with the right index finger (to the right direction) that begins on the third person in line 007 and stops at the untimed pause.

And that inserted commentary refers to important directions for maneuvering locations in the house. Notice that Black repeats the *in*-locative in both lines 007 and 008, an augury of its looming importance. According to Black, the significance of Bowman's current position involves its location relative to the escape route out of the home and into the parking lot (line 011): a room *right next to the back door*. However, while the place adverbs (*here* in line 010) and deictic particles (*next to* in line 011) refer to the location of the *back door* (line 011) they do not function as vividly, economically, and efficiently as gesture in providing the jury with changing directions in the diagram.

As mentioned above, in line 007 Black begins to trace Bowman's movements in the home but interrupts the gesture when he aborts the projected correlative. That is, on the third person he starts to trace (with the right index finger) toward the location of the back door but aborts the gesture at parenthetical onset. Moreover, when he inserts the parenthetical into the host utterance from lines 007–011, his recalibration of speech co-occurs with a change in gesture, from tracing to right-hand index finger beats, and this demonstrates our prior point about speech-gesture synchronization. More explicitly, Black starts to trace with the right finger but aborts that activity when he aborts speech in line 007, then restarts the tracing motion when he recycles the correlative conjunction in line 012. On the *in*-locative and demonstrative complement (*this room*), he employs three index finger beats to foreground Bowman's location in the home while gazing at the diagram (line 007). The beats consist of the right arm bent at the elbow with the index finger extended and other fingers curled under the hand/thumb in an up-down motion. From lines 006–011 his right index finger continuously touches the relevant location on the diagram. In line 008, he shifts gaze to the jury and then delivers two more beats over the third person and spatial locative (*she's in*) (See Figure 3.8 below). Beating gestures foreground the *in*-locative and spatial image of enclosure but, most impressively, mark or beat time: a metaphorical duration of time through duration of beats. That is, the beats do not just occur anywhere in the utterance but land on the spatial locatives, illustrating temporal duration visually. The location of the door shows Bowman has the opportunity to leave. The temporal duration conveyed through the beats shows she has the time to leave. Together they reveal that despite her location and time she does not take the opportunity to escape her assailant. Thus beats not only orchestrate the rhythmic pulse of speech and foreground the *in*-locative, they also beat out duration of time as a temporal metaphor to indicate the amount of time Bowman has to leave the house – an opportunity especially significant after a sexual assault.

Figure 3.8: (line 008 *in*).

Most importantly, Black never mentions what room Bowman is in, only *this room in the home*. After another beat on the demonstrative in line 009 (*this conversation*) he returns gaze to the diagram and traces movement – a move Bowman should have taken – from the current room location to the *back door* and back: Bowman's location relative to the escape route out of the house. In the process, Black appears to imply that it would be odd for an alleged victim to have a "conversation" with the defendant after he had just sexually assaulted her.

In theoretical terms, tracing is an iconic gesture superimposed on an iconic representation (the diagram): a double icon. According to Goodwin (2003: 229) "Pointing gestures can trace the shape of what is being pointed at, and thus superimpose an iconic display on a deictic point within the performance of a single gesture." His tracing involves a directional finger movement against the surface of the diagram while gazing at the gesture, providing a precise icon of movement along a specific route. Just as crucial, Black appears deeply engrossed in the delicate tracing gesture on the diagram, reifying and objectifying it as a source of legitimate knowledge.

In terms of our prior points on the impact and retention of evidence (Mauet 2017: 95), finger tracing enhances spatial recognition of route alterations, guiding jurors to crucial information (such as the location of the back door), while providing a visual record of movement in the environment, much more effectively than speech alone. What is crucial for the jury is this. Tracing makes it easier to process directional information and devote a focus of attention to directional alterations on the diagram (Wermeskerken et al 2016; Ginns et al 2016).

Using Black's body facing the diagram as the deictic center, we see his index finger traces an inch or so to the right and back. Once again, without the tracing gesture the jurors would have no indication of Bowman's location relative to the *back door* and, more importantly, how close that escape route is relative to her current – her *stays* – location. It gives the jury a precise icon of movement along a specific route and the locational field in which she is situated. Although speech marks the location of the current room relative to the *back door* it does not capture one room's location relative to the other and the change in direction for the figure to get from one room to the other and back. Tracing is a much more powerful representation than mere pointing. While pointing gestures can show deictic location in space, they cannot demonstrate or illustrate as effectively as tracing, directional alterations of movements or flow of movement. In so doing, we see a vivid illustration of how materiality, gesture and speech mutually elaborate one another in a contingent display of evidentiary knowledge.

As an ideology of communicative practice – of power – Black frames and naturalizes Bowman's location and movement in the house as an agentive strategy rather than, for example, rape trauma syndrome and the disorientation that accompanies it (acute characteristics such as disorganized thinking, confusion, and shock that would diminish her agency). She chooses to *stay* rather than flee, a choice we will explore in more detail in the next section. Of course, his framing strategy assumes Bowman knows where the back door is and that it leads to the parking lot. Most importantly, the jury gets to experience the route also, as Black instructs them on how they would have (commonsensically) taken the back door out to the parking lot. That is to say, he allows the jury to become participants actually moving in the then-and-there house. In Goodwin's (1994) terms, framing practices guide "seeing" through tracing on the exhibit, organizing instructions for its interpretation and authorization of its evidential facticity.

Although the parenthetical is often referred to as an "orphan" or "unattached" constituent that does not mean it is unimportant as a communicative-legal strategy. In the case here, Black needs to set up the relevance of Bowman's movements beginning in line 012. The tracing from lines 012–16 will not make sense as inconsistent with post-incident behavior unless he sets it up or contextualizes it with the parenthetical first: a precise sequence of motion. Thus the parenthetical clause contributes to the sense and significance of the forthcoming tracing motions in the various locations of the house. He has to show the importance of back door location before recycling the aborted correlative in line 012, as we will see next.

3.3.3 Rummages around the property: Tracing motion events (lines 012–016)

In line 012 Black reveals why the inserted comment is so vital to his communicative strategy. The parenthetical consists of preliminary work produced with an eye toward the motion event representation in lines 012–016. More specifically, he recycles and resumes the correlative and tracing gesture that accompanies it as the first part of an oppositional structure: *She not only does not go out the back door.* He gazes at the jury in utterance initial position, produces a lateral headshake that begins on the *do*-Auxiliary (and ends on *back*), and traces his right index finger (with the right arm bent at the elbow) from Bowman's current location to the *back door* and back. Indeed, the headshake, along with uncontracted *does not,* represents a negative metapragmatic comment on her directional choice and its implications for her allegations.

Just as important, he positions the body at an oblique angle or slight torque/slant with the right shoulder facing the diagram and the left shoulder toward the jury, keeping multiple states of involvement in play simultaneously.

In line 013, he starts the second part of the opposition using an event structure listing of directional motion verbs and path prepositions. First, Black recruits the directed motion verb *go,* path preposition *through* (that not only encodes a three-dimensional space but also denotes purposeful activity), and path complement *the kitchen.* Second (in line 014), he repeats the path preposition and adds the *dining room* complement as Bowman continues a trajectory toward an end point. And, third, he concludes the listing opposition with the boundary-crossing path preposition to encode her reaching the goal: *into the living room.* Indeed, *into* is a powerful word choice because it is not only a boundary-crossing path preposition, but, like *in,* it also conceptualizes the landmark metaphorically as a container (*the living room*). So, in parallel with *stay in the house* (which is also a container-type landmark) *into the living room* invokes an image schema that undergoes a transformation to include a representation of movement from outside the container to inside it.

Put another way, the correlative conjunction marks a parallel structure in progress (the first part), the second part occurs in the listing of goal paths that provide a striking rhythmic contrast (*through the kitchen* etc), and the third evokes the rhythmic resolution with the search verb (*rummages around the property*). As it reaches a poetic crescendo, metricalized integration of co-speech gestures, speech and diagram furnishes a powerful argument about evidence and witness credibility.

However, space, direction, and motion are encoded through not just speech but, even more vividly, co-speech tracing gestures. In line 013, Black traces downward with the index finger several inches on the diagram (See Figure 3.9 below). Moreover, the gesture not only physically touches the diagram during the path-evoked motion but also registers an audible scraping sound or acoustic effect over the route; that is, the jury not only sees but hears the tracing gesture.[10] Such an audible effect – a deeper trace on the surface – further animates a deeper movement into the home and away from "safety." In line 014, Black alters the direction of the route with a second tracing gesture to the left several inches (or if we use Bowman as the deictic center figure, she changes direction to the right), moving from *the kitchen* to *dining room*. And, finally, Bowman reaches her route destination, *into the living room*, with a third tracing gesture that moves several inches further to the right. This trace represents the final increment to Bowman's progressive movements and departs slightly from the form of the prior two gestures (See Figure 3.10 below). First, Black's index finger lifts slightly off the diagram during the tracing motion and lands forcefully on the *living room* location. And, second, his finger stays on the adverbial (for over a second) as a post-stroke hold for further emphasis. For still further emphasis, the goal paths in lines 013 and 014 occur with marked stress, building up rhythmic intensity and momentum toward the final location crescendo (*into the living room*), which not only occurs with marked stress and gestural intensity but noticeably increased volume as well. Just as crucial, Black marks Bowman's course with both his tracing finger and eyes. He gazes at the tracing finger motion as it moves along the path direction and shifts in direction, finally shifting gaze from the diagram to jurors on the final path complement (*into the living room*).

Figure 3.9: (line 013 *through*).

---

10 Gilbert and Matoesian (2021: 97) refer to these as *acoustic gestures*, such as the finger snap, scrapping sounds against an object etc.

Figure 3.10: (line 015 *into*).

Let us consider, in more detail, the delicate relationship between *through* and *into*, on the one hand, and tracing gestures on the other. *Through* conveys a sense of continuity in three dimensional space or means to an end; it designates "a place as point on a path" (Taylor 1993: 163). As mentioned, *into* encodes a boundary-crossing path and sense of completion. The tracing on *through* consists of gestures flush on the surface of the diagram. The tracing gesture on *into*, however, begins flush on the surface but then moves up and off the diagram slightly and concludes with a downward "landing" motion on the final destination landmark (*into the living room*). This not only shows speech-gesture synchronization but also how the *form* of speech-gesture synchronization shifts to coordinate a more nuanced meaning than either could impart alone.

To develop our points on co-speech tracing gestures, speech alone does not show where the kitchen, dining room, and living room are located relative to one another; nor does it show directional shifts in movement from one location to the other – at least not as effectively and economically as co-speech tracing gestures. As Klippel et al (2013: 116) mention, the more complex the directional movements, "the more verbose is the description," descriptions requiring numerous qualitative options in dense combinations like hedges, projective terms and degrees for adequate interpretation. In a similar vein, Goodwin (2003: 230) notes, "tracing provides a way of indicating precise information about what is pointed at . . . that would be difficult to specify through language alone."

In sum, Black's polyrhythmic and multidimensional poetic format is organized as follows: (1) The contrast generates a violation of normative expectations by encoding paths followed by the figure (Bowman), even more so than *staying* in the room; (2) the parallel listing of motion paths foregrounds and amplifies the violations in a cumulative format; (3) gaze keeps multiple forms of involvement in play simultaneously; and (4) tracing furnishes ease of processing and visualizes directional shifts, much more so than could be achieved through speech alone (Hu et al 2015; Hu et al 2014). Together each modality mutually elaborates the other to bring into focus sharply defined zones of relevance in an otherwise dispersed diagram of regions to authorize the facticity of his account.

And that poetic format plays a major substantive role driving the transformation from an opening statement into an argument about witness and evidence credibility. Black presents the diagram as an objective medium for transmitting an outline of information, a perceptually neutral field of locations and movements in the house. As we have seen it is quite a lot more than that.

First, the incremental progression of movements and route alterations in the house display Bowman moving toward a goal, showing that she was not fearful, and such post-incident behavior is inconsistent, according to Black, with having been raped just moments before. Just as important, the historical present (in the verbs) provides a sense of deictic immediacy for the jury to walk along with her, and that provides a more vivid way for them to experience her inconsistent behavior. In a powerful ascription of agentive identity, Black shows how, after the discussion mentioned in line 001, Bowman leaves the room and moves through the various locations unilaterally, without external motivation (from Smith for example). Second, in lines 001–005 Black shows what Bowman does not do: *flee the house.* Instead (in lines 012–016) she moves through the various rooms to the end goal of the *living room,* where she engages in the bizarre behavior of *rummaging around the property.* Black's rhythmic resolution culminates in the search verb (or what Levin 1993: 71, 199, calls a "rummage verb"), which evokes a negative connotation of "unsystematic rifling through a mass of goods," which also indicates she has plenty of time (which works with "stays" in this regard). And, third, Black's motion and path framing militates against and erases other relevant explanations for Bowman's behavior, such as rape trauma syndrome. Her movements rather than Smith's force become the focus of attention.

3.4 Summary

While orthodox research on the encoding of motion events focuses on semantic-grammatical relations, the current study shows the analytic pay-off for inspecting the mutual elaboration among language, gesture and material artifacts in the construction of space, path, and motion. More explicitly, deictic forms in motion events become crystalized relative to speech through gestures and the diagram. We have seen how tracing gestures on the surface of the diagram shape a dense field of diffuse locations into concentrated zones of evidentiary relevance – a strategic positioning of demonstrative evidence to transform opening statement into an argument about evidence and witness credibility. We also saw how an ensemble of beat gestures produces a rhythmically subtle framing strategy toward the material object (the exhibit), transforming a static non-motion verb (*stay*) into a motion verb.

And that argument involves the "tour" of the home. Most significantly, the fluid movement of tracing on the diagram encodes the motion event path in a most vivid and engaging representation (via the stressed *throughs* for example), a representation that would be left underspecified in speech alone. Black's performance places the jury in the tour and lets them experience the tacit (not explicit) inconsistencies he shapes: Bowman's departure from normatively gendered expectations. As we have seen, the longer he gives the "tour" the longer the jury takes part in it and thus the more glaring the inconsistencies appear, as he bleaches other possibilities like rape trauma syndrome. Of course, while on the surface she appears to violate the gender order, on a deeper level those violations involve a strategically driven system of sociolegal hegemony to naturalize them.

In more theoretical terms, the construction of rape's legal facticity trades on this multimodal infrastructure of speech, gesture, and exhibit to generate ideological fields that shape inconsistency, credibility, and evidence in taken-for-granted ways. Forms of power like framing (or setting the agenda) and sociolegal hegemony are strategically driven through this integrated semiotics of multimodal action, a poetic organization that shapes ideological fields into perceptually salient and concentrated zones of evidential relevance. While the law defines inconsistency as a generic juxtaposition of contradictory or incompatible facts of evidence (a rational system of logic) we have seen here how it represents a linguistic ideology for constructing the normative organization of gender identities in the trial (a logic of power for the defense).

In chapter 4, we explore another – indeed rather novel – dimension of opening statement, how it summons an intense emotional outpouring of not just personal but collective grief, once again revealing how opening involves much more than a mere outline of what the forthcoming evidence will show in the case.

4 Historical voices, collective memory, and interdiscursive trauma in the legal order

On November 22, 2013 thousands of people gathered in Dealey Plaza in Dallas, Texas to commemorate the 50th anniversary of President John F. Kennedy's assassination. Bells tolled at 12:30pm Central Time (the time shots rang out on November 22, 1963), and parts of the speech he was to give that day were enshrined in a concrete memorial to commemorate one of the most beloved figures in American history. On June 6, 2018 at Arlington National Cemetery in Arlington, Virginia another assembly of thousands gathered for the 50th anniversary of the assassination of Robert F. Kennedy, with guest speaker President Bill Clinton delivering the main address. At the end of the address, Clinton stated that Robert Kennedy "still lives in millions of hearts and millions of minds who wish and hope for a better world." Indeed, for millions of Americans the assassinations of RFK and JFK were traumatic events of unparalleled historical importance – of collective shock and grief. As Eyerman (2011:157) notes, these events robbed both the "person and collectivity of their potential for greatness." We lost who they were, what they could have been; and who we were and what we could have been as a nation. In *The Last Campaign,* Thurston Clarke (2008: 10) wrote that if RFK had lived and won the 1968 Presidential election, it would have saved the lives of two million Vietnamese and twenty thousand Americans. There would have been no war on college campuses, riots at the Democratic Convention, and Watergate. Such sacred memories and the commemorative events that orchestrate them shape our interpretation of not only past but also present and future possibilities, intensifying collective sentiments, reaffirming cultural identity, and linking individuals to the group.

As important as these formal and planned commemorative events were, there was another commemorative ritual over two decades earlier, one that was unplanned and certainly unanticipated yet more widely seen and significant for the Kennedy family. On December 2, 1991, opening statements began in the widely publicized rape trial of William Kennedy Smith – the nephew of JFK and RFK. Although the adversarial trial should consist of the facts in issue, on occasion it is not always or even primarily about that. At certain moments, the trial may be less about evidence than a form of commemorative symbolism to evoke collective memory and resurrect national trauma about events of great historical significance. In the case here, trial participants transform legal context into a sacred ritual of collective memory, if only briefly, into an allusive yet emotionally charged representation of historical events and co-construction of socio-legal identity.

4.1 What is collective memory?

Maurice Halbwachs (1992) introduced the concept of collective memory to highlight the social basis of memory, to explain how social practices breath historical continuity into cultural life and reproduce space-time cohesion in society. Building on Durkheim, he sought to understand how collective memory(s) functions to maintain group solidarity, reproduce social order, and frame cultural identities – a social fact that generates moving images at the intersection of microcosmic order and historical context. As Schwartz (2007: 588) puts it, collective memory refers to the "distribution throughout society of what individuals know, believe, and feel about past events and persons, how they morally judge them, how closely they identify with them, and how much they are inspired by them as models for their conduct and identity." In so doing, collective memory reproduces a sense of continuity between the past and present – reconstructions of the past that provide order and coherence to events of deep emotional significance for a specific age cohort. Indeed, the past is kept alive through commemorative symbolism like ritual oratory, eulogy, conversation, archives, and narrative (Schwartz 1982), or as French (2012: 337) puts it, through "publicly circulating signs," including embodied practices.

In this chapter, rather than consider collective memory as a static social fact we follow French (2012) to show how it circulates in the poetic details of multimodal conduct in court – in the integration of gesture, language, and emotion.[11] We investigate how sociocultural memories are resurrected, reproduced, and reconfigured in the present context to accomplish distinct legal evidential work – how ritual oratory generates allusive images of political assassinations and transforms JFK and RFK into sacred objects of collective memory. We examine how attorneys and witness co-construct an emergent legal space for a commemoration of collective sentiments and cultural trauma, solemn worship of the god-like figures. The collective memories that emerge in opening statement and witness questioning co-construct a socio-legal identity of shared victimhood as a strategy for framing evidence and testimony. Just as important, collective memories are not merely "transmitted" via the referential content of historical events; rather the poetic metricalization and affective form of multimodal performance itself steers our interpretation of collective sentiments and shapes the relevant trial issues embedded in it. That is to say, collective memories involve much more than the mere interdiscursive transmission and selective reconstruction of referential content from the past. Multi-

---

11 As Bietti (2010: 520) (see also Wodak and De Cillia 2007) notes, the production of the past in the present is always a selective and situated reconstruction not some objective representation of historical events.

modal performance shapes the sociocultural interplay of macrocosmic past and micro interactional present with an eye toward the evidential task at hand.

4.2 Gesture in microcosmic action

As mentioned in chapter 1, unlike iconic and metaphoric gestures that represent images of some object or action, beat gestures are, according to McNeill, non-imagistic. They orchestrate the rhythms of speech and highlight significant information, landing on stressed syllables to visually parse speech into prominent segments to increase the persuasiveness of the message. That is, beat gestures (in the orthodox view) do not depict an object of reference but, like an orchestra conductor's baton, beat out rhythm and provide visual structure that mark specific parts of speech as significant (Streeck 2008), representing the acoustics of speech in movement.

While most gesture studies focus on representational gestures or gestures that depict semantic or propositional content, few in-depth studies have been conducted on beats, perhaps because they are thought to merely beat out "musical time" rather than convey substantive information (McNeill 1992: 15). But as we saw previously (and as we will see in more detail in chapter 5) beats perform crucial functions in persuasive oratory (Ferre 2011), far beyond just beating out time.

Speech and beat gestures integrate with other forms of bodily conduct such as gaze and facial expressions in the production of persuasive oratory. We consider the courtroom, at least on occasion, as a poetically dense ritual that represents group identity and ignites collective sentiments, an interdiscursively driven aesthetic performance linking macro historical collectivities to micro action.[12]

This chapter is organized as follows. After a brief introduction of the data, the ensuing section examines the opening statement of the defense attorney, a poetic oratory that consists of historically circulating representations of and allusions to political assassinations. The interplay of microcosmic practices and historical meaning contextualizes a legal space for co-victim identities in an affect-laden ritual of collective grief and national trauma. The attorney transmits historical trauma into the present and transforms it into a solemn socio-legal strategy through a multimodal integration of speech, gesture, and affect.

The final section analyzes direct and cross-examination of Senator Edward Kennedy and how he invokes allusive memories of his brother, Robert Kennedy, (and other family tragedies) during the night of the assassination. In so doing, he

---

[12] Interdiscursivity refers to the historical circulation of prior utterances, genres, and voices and their dynamic interplay with present performance.

not only shapes legal-evidential identity but also creates interdiscursive links with RFK's announcement in Indianapolis, Indiana of the assassination of Martin Luther King (April 4, 1968) and the eulogy he gave at RFK's funeral in June 1968. But this is not a sole speaker's narrative. Senator Kennedy's solemn performance emerges as an interactive co-construction with attorneys in direct and cross examination, a ritual performance that creates dense laminations of interdiscursive symbolism in an emotional mapping of denotational text to interactional function. The sacred objects of commemorative ritual that materialize in the trial allow jurors to step into history and connect with the Senator's reenactment of national tragedy and personal grief – backgrounding, perhaps even erasing, the trauma suffered by the rape victim in the process.

4.3 A "strange" feature of the legal case

A strange feature of the case occurred when the prosecution, led by Moira Lasch, announced they would call Senator Kennedy during direct examination, since he was at the estate that evening. However, the purpose of her questioning (or why she even called the Senator in the first place) was unclear, especially given that the defense most likely would have never called him because of the possible liability of his tarnished moral character. In an interview after the trial, defense attorney Roy Black told Matoesian (see Matoesian 2001) that calling Senator Kennedy was "one of the biggest mistakes of the prosecution" and that "his testimony changed the atmosphere of the case." Moreover, by calling the Senator on direct, the prosecution opened the door for Black to discuss Senator Kennedy during opening statement and then conduct his own friendly cross-examination after direct. Black stated in the interview: "He came in and sat down. Here's a guy who's a living legend. Whether you like Ted Kennedy or not, this guy's a piece of living American history. He comes down and sits five feet from the jury and starts talking to them. That was a fascinating piece of the case at least psychologically." In what follows, we turn to a portion of Black's opening statement in which he introduces Senator Kennedy to the jurors.

Example 1 Defense Attorney Roy Black Opening Statement 3:22–5:22

((lines 1–13 vertical oblique open hand beats). B=Beat))
01 I wanted to start out by::: talking to you a
   (B) (B)   (BHE) (BDS) (HE=high elevation upstroke;
                                     DS=Downstroke)
02 little bit about Ted Kennedy (1.2)
   (B) (B)   (B)

03 I know how   (0.5) people feel about Ted Kennedy
      (BUS) (BDS)   ("empty" beat up and downstroke + hold)
(BUS=Beat Upstroke; BDS=Beat Downstroke)
04 and I think it's- (.) important
            (B)   (B)
05 to discuss *why he was at the home* (.) ((staccato on each word))
         (B)   (B)   (B) (B)   (B)
06 that particular weekend (1.4)
       (BUS)   (BDS+Hold)
07 What many people may not know (0.9)
      (BHE)   (B)   (B) (B)
08 is that there are twenty-one
         (B)   (B)
09 children (0.9) in the Kennedy family (0.4)   who
         (B+hold)   (B+hold) (B)
10 do not have a ***father*** (1.3) For those
     (B)   (B)   (B+hold)   (B)
11 who do not have a father (.) Ted Kennedy (.) is
          (B)   (B)   (B)
12 their surrogate father (1.6)
                 (B+hold)
13 Whatever it may be (0.9)
         (B)
((lines 14–22 vertical forefinger beats with the lower fingers curled in toward palm))
((speaker shifts gaze direction to align with direction of beats))
((Noticeably marked beats))
14 whether graduation
              (BHEUS)=(Beat with high elevation upstroke)
15 from *high school* (0.9)
          (BDS+Hold)
          ((Center position))
16 from   *college* (0.9)
     (BUS)   (BDS+Hold)
                    (B to right)
17 from   *law school* (0.6)
     (BUS)   (BDS+Hold)
                    (B further to right)
18 from   *medical school* (0.6)
     (BUS)   (BDS+Hold)
                    (B back to left)
19 whether it be
            (BUS)
20 a *wedding* (0.6)   or   a christening (0.8)
      (BDS+Hold)   (BUS)   (BDS+Hold)
      (B center)   (B to left)
((the beats (B1) in 21–22 begin with a high elevation beat that progresses

downward incrementally till the end of the intonation unit, then recycle
   up again and back down 3x recursively)) ((noticeably marked pattern))
21 those children (.) have Ted Kennedy there (0.5)
     (B1)   (B2)   (B1)   (B2   (B3)
22 as their father (2.0)
     (B1)   (B2+Hold)
((several lines omitted))
((Open Palm Vertical/Oblique with Right Hand Beats 23–27))
23 In August of Nineteen-Ninety Steven Smith (0.7)
         (B)   (B)   (B)
24 Will's father died (1.0) after
                             [((extends open palm to jury box))
25 a long illness (0.9)
26 Ted Kennedy was with him (0.8) throughout his treatment (0.7)
     (B)   (B)   (BHE)   (BDS+Hold)
                                 ((Noticeably marked up and downstrokes))
27 He was with him when he had his   operation (1.0)
    (BHEUS)   (BDS)   (BHEUS)   (BDS+hold)
                                 ((Noticeably marked up and downstrokes))
((Index Finger Downward Beats with Right Hand 28–29))
28 He was with him when he died (0.9)
    (BIF)   (BIF)
    (Right)   to   (Left)
29 He gave the   eulogy at his funeral (1.7)
    (BIF)   (BIF)
    (Right)   to (Left)
((several lines omitted))
(Open Palm Vertical/Oblique Beats with Right Hand 30–31)
30 Ted Kennedy *went* (.) every chance he could
       (B)   (B)   (B)   (B)
31 get (.) to be with the Smith family (0.7)
     (B)   (B)
32 because once again (1.5)
   (palm down and flip to vertical/oscillating gesture)
(Lateral Open Palm Prone Beats in 33–34 over "another" + "widow"/ "case")
33 another   widow in the family (1.4)
    (BUS)   (BDS+Hold)
    (left) to (right lateral beat)
34 another   case (.)
    (BUS)   (BDS)
    (right) to (left lateral beat)
(Lateral/Oblique Open Palm Beats with Right Hand in 35)
35 of children   without   a father (3.5)
     (B further to left from 34)   (B to right)   (B further right+Hold)

4.4 Poetic oratory as microcosmic ritual

Cultural trauma refers to the communication of affect or emotion – the "language of affect" – and represents a key element of collective memory (Smelser 2004: 40–41; see also Alexander 2012: 2). Eyerman (2012, 2015) notes how emotions are central to the concept of cultural trauma: a deeply moving emotional event or shock to the collective conscious, such as the assassinations of JFK, RFK, and MLK. As he (Eyerman 2011: 25) puts it: "cultural traumas are also a response to deeply felt emotions." Using speech act theory, Alexander (2012: 16) refers to the persuasive "performance of trauma," a cultural process of "speeches, rituals . . . storytelling . . ." that ignites powerful emotions among members of the collectivity in response to disruptions in the social order (Alexander 2012: 4).

Although in general agreement with Alexander's formulation, we need to expand on his position and pose the question: what does the language of affect look like in the concrete details of situated conduct? Consider a quote from defense attorney Roy Black in the interview Matoesian conducted with him in 1994. Matoesian asked Black how he would have defended the boxer Mike Tyson, who had recently been convicted of rape. In response to a comment that his client was well educated and articulate compared to Tyson, Black mentioned that education had nothing to do with it. Black thought that Tyson's lawyers failed to show his emotional side. He went on to say:

> I can take some guy off the street and ask him to get in and talk about something he passionately believes in and he'll be eloquent . . . and you can get the most accomplished orator in the world, Mario Cuomo or Jesse Jackson, and you get them talking and if they don't have the passion for what they're talking about they're not going to convince anybody . . . A lot has to do if the jury perceives you're human, that you're being truthful, that you're disclosing who you are to them, that you're telling it *from the heart*. [our emphasis]

In their seminal work on affect in language, Ochs and Schieffelin (1989) argued that, similar to Black, "Language has a heart" or how feelings and similar so-called mental predicates are encoded in language. (1989: 9): "Interlocuters need to know not only what predication a speaker is making; they need to know as well the affective orientation the speaker is presenting with regard to that particular predication." Since then work in linguistics and related disciplines has shown not only that "language has a heart" but that gesture, facial expression, and bodily movement integrate with language in the production of affective or emotional discourse (Wilce 2009; Fleming and Lempert 2014).

How does multimodal conduct encode an affective stance relating to collective memory and the cultural trauma embedded in it? What would the cultural trauma of affect look like in the concrete details of microcosmic action? Rather than using interviews, aggregate data, or the various "proxies" of emotion like film, books, music or anecdotal glosses or vernacular accounts, we investigate multimodal affective practices in the unfolding rhythms of interactive context. How does collective memory and cultural trauma inhere in the microcosmic practices of the law? Here we show the dynamic relationship between collective memory and the microcosmic features of multimodal conduct: how the trial becomes a site of public mourning and a solemn ritual of collective grief.[13]

In lines 01–03, defense attorney Black begins a narrative about Senator Kennedy (*I want to start out by talking to you a little bit about Ted Kennedy*), but then provides a not too subtle insinuation about his tarnished reputation: *I know how people feel about Ted Kennedy*. That is, most people know Kennedy through negative publicity from the incident at Chappaquiddick, his excessive drinking, womanizing, and so on, flaws that led Newsweek Magazine (12/8/91) to label him as the "living symbol of the family's flaws." However, as it turns out that assumption serves as the moral platform for the forthcoming positive (and unexpected) contrast component that begins in line 07 (*What many people may not know . . .*).

And the narrative involves more than just speech. In line 01, Black mobilizes a series of beats to foreground the points about Senator Kennedy. Of special interest is vowel lengthening on *by:::* that co-occurs with a hold on the high elevation upstroke that waits for the verb *talking*. In line 02, he produces another series of beats with an open-palm supine gesture on the 1.2 second pause to offer information to the jury.[14] Of special interest is the gesture and pause in line 03, which demonstrate how gestural form itself conveys meaning beyond the substance or topic of talk. Here, Black produces a beat upstroke at turn onset and a downstroke that lands on *how* prior to the 0.5 second pause. During the pause he executes another beat but this time without a corresponding lexical affiliate: an "empty" gesture. The downstroke includes a "hold" till clause onset in line 04. In this instance, Black mobilizes the empty gesture in concert with the hold+pause to signal an embodied sense of reluctance toward the forthcoming negative allusion about Ted Kennedy. Once again, the marked embodied hedge – the form of the gesture – functions as a display of reluctance concerning unflattering characterizations about the Senator.

In lines 04–07, Black begins to explain the Senator's presence *at the home that particular weekend*. In terms of the co-temporal alignment between speech and gesture, the upstroke after *discuss* is held during the .4-second pause before the

---

13 Still more theoretically, we take up the empirical challenge put forward by Wilce (2009: 190) when he recommends that: "Ethnographies of local linguistic emotional practices ought to take account of the positioning of those practices vis-à-vis relevant histories."
14 According to Kendon (2004) speakers employ the open palm supine or palm up gesture to offer something to or request something from the interlocutor(s).

beat down stroke resumes (allowing speech to catch-up with the gesture). After that, notice, in particular, the staccato emphasis over the embedded *wh*-question is accompanied by several open-palm vertical beats over the clause. The slow staccato speech keeps alignment with the beats and maintains the co-temporal integrity of gesture-speech ensemble. Once again, the co-occurring beats add an emotional charge to Black's ritual oratory and foreground the main points in his narrative (see Figures 4.1 and 4.2 below). And a good deal of that affect is organized around the pauses and corresponding gesture holds in lines 09, 10, and 12 that lead up to the allusive crescendo about the assassinations of J.F.K. and R.F.K.

Figure 4.1: line 07 (vertical open palm beat upstroke on *What*).

Figure 4.2: line 07 (vertical/oblique open palm beat downstroke on *people*).

Under auspices of merely telling the jury *why he was at the home that particular weekend*, Black produces an emotionally moving transformation from a profane to sacred performance that has little to do with the denotational text above. As the sequence unfolds, he executes a thoroughly unveiled allusion about the assassinations of J.F.K. and R.F.K. in the most emotionally stirring representation possible (lines 08–10): *twenty-one children in the Kennedy family who do not have a father.*[15] In so doing, he not only delivers an affect laden circulation of collective trauma about assassination but also conveys the personal trauma suffered by their fatherless children, a trauma that thrusts Ted Kennedy into the virtuous and gallant role of *surrogate father* (line 11). Black designs person reference with an eye toward collective memory rather than denotational or kinship reference: a reference that does considerable interactional work in creating a positive, perhaps unexpected, socio-legal identity for the Senator and recruiting that identity to contextualize the forthcoming parallel structure.

And in lines 13–22, he proceeds to just that. He itemizes or lists the duties of the surrogate role incumbent with a highly affective and perceptually salient form of poetic patterning. As mentioned previously, poetic function is the way language draws attention to itself (Jakobson 1960), and parallelism or repetition with variation is the most culturally salient form of aesthetic foregrounding. As Wilce (2009), Fleming and Lempert (2014) and Stasch (2011) have indicated, ritual oratory exploits parallelism as cultural performance to move the audience and invoke the sacredness of the message (see also Tavarez 2014). In a similar vein, Besnier (1990: 426) states: "parallelism represents the convergence of affective and poetic dimensions of language. It represents the primary device of poetry because it possesses an emotional effect."

Black's metrically infused performance begins with the embedded *whether*-clause (which precedes items in a series) consisting of the Prep + NP listing of ceremonies the surrogate father attends: *from high school, from college, from law school, from medical school.* In a strikingly nimble touch of rhythmic agility, he then reconfigures it into the subjunctive (*whether it be* line 19) platform to resume the repetitive progression: *a wedding* or *a christening.* He packages his narrative in metrically driven and emotionally charged rhythms that guide interpretation and provide a sense of natural cohesion to the listing items (even though most of the "children" are adults).

But Black's emotionally evocative pattern consists of quite more than measured repetition in the aesthetic text. He integrates language and gestural metricalization to produce an affective form of multimodal poetics or cross-modal parallel-

---

15 *Father* occurs with both increased loudness and stress for additional emphasis.

ism, a poetically dense oratory that links ritual to historical context. In light of our previous discussion of the co-temporal coordination between gesture and speech, let us reconsider the poetic function of language. Black's improvisational oratory integrates both verbal and visual modalities to assemble a polyrhythmic and multidimensional performance – a dense weave of poetic patterning that calibrates an allusive-emotional ritual of collective memory (and one that bestows a measure of distinction upon the Senator). That is, the poetic function inheres not only in language but through the integration of language and gesture in the cascading polyrhythms of ritual oratory.

In line 13, the structure of Black's beats undergoes a major transformation to develop the points about Senator Kennedy. The beats shift from oblique open palm vertical movements that accompany allusive images of assassinations to forefinger beats organized around the itemized listing of surrogate role activities (the index finger extended with other fingers curled beneath the thumb and flexed toward the palm – see Figures 4.3–4.5 below). In this case, the beat upstroke occurs on the preposition and the down stroke on the NP, with a hold co-occurring on the pause for added poetic affect and emphasis: gaining momentum in the process (*from high school 0.9, from college 0.9, from law school 0.6, from medical school 0.6*). Why the shift in beat structure? The oblique open palm beats convey a general idea; index finger beats itemize specific facts for accumulation. Together, their variation captures form-function relations in the unfolding narrative.

Figure 4.3: line 14 (*whether*).

Let us look at the depth of cross-modal patterning or multimodal poetics in more detail, the integration of modal resources from lines 13 to 22. Once again, each beat has an upstroke on the preposition and downstroke on the NP, and each item in the

Figure 4.4: line 15 (lateral beat on *high school*).

Figure 4.5: line 17 (lateral beat with gaze shift on *law school*).

parallel structure ends with a pause, co-occurring hold on the beat down stroke, and stress on the NP. In addition, each beat consists of elbow beats with the right hand, and each *whether*-clause begins with high elevation in the center position (or high elevation in the beat upstroke marks the *whether*-clause in lines 14 and 19). Most impressively, each repetitive beat consists of not just vertical up-down motions but lateral (with a low-arc trajectory) to-fro progressions on each ceremonial increment, aligning gaze with beat direction to engage jurors (that is, the beats and gaze move from left to right in lines 16 and 17 and then back to the left in line 18 and finishing back to the center in line 20). These equi-measured beats segment ceremonial increments to visualize the accumulation of listing items, once again demonstrating how gestural form signals metaphoric meanings in addition to rhythmic foregrounding.[16] And, finally, the interlocking rhythms reach the crescendo in lines 21–22: *Whatever it may be ... those children have Ted Kennedy there as their father.*

Thus the unfolding multimodal repetition is not just polyrhythmic (speech, gesture, stress, and gaze) but multidirectional (vertical and lateral gestures) as well. Black recruits a densely layered polyrhythmic and multidimensional progression of allusive images to impart collective memories of not just the assassinations of JFK and RFK but the fatherless children they left behind. And there is still more historical tragedy to come.

In line 23, Black finally reaches the explanation for Senator Kennedy's presence at the estate that Easter. He adds the death of Steve Smith (father of William Kennedy Smith and husband of J.F.K./R.F.K.'s sister, Jean Smith) on to the allusive images of J.F.K., R.F.K. and their fatherless children, once again using a poetic litany of elements with the Senator as the subject (in lines 26–29: *Ted Kennedy was with him throughout his treatment; He was with when he had his operation; He was with him when he died*). More technically, Black employs the 3rd person (he) + Be + 3rd person object (him) and relative to construct a loyal and altruistic identity. For further emphasis, he uses noticeably marked silence at the end of each item in the parallel structure (0.7 second pause in line 26, 1.0 in line 27, 0.9 in line 28, and 1.7 in line 29), a pattern that culminates with the off-kilter and slightly dissonant flourish in line 29: *He gave the eulogy at his funeral.* More provocatively, Black enlists Steve Smith as a strategic feint to evoke allusive images of J.F.K., R.F.K., and their fatherless children.

And those images unfold in and through the repetition infused integration of speech and gesture. Once again, Black mobilizes a series of oblique open palm beats to accompany speech and foreground the main points in his argument in lines 26–27. Of special interest is that the beats in line 26 occur with noticeably marked up and downstrokes, with the beat that accompanies *throughout* possessing extremely high elevation on the upstroke and deep bottom on the downstroke as a metaphor for the lengthy duration of time the Senator spent with his ailing brother-in-law (see Figures 4.6 and 4.7 below). Although, as mentioned earlier, McNeill (2005) claimed that beats are not imagistic, merely beating out time, the beat in line 26 not only functions to orchestrate the rhythms of speech and foreground key points of the message, but also works simultaneously as a symbolic metaphor for the duration of time.

While the beats in 26–27 consist of the oblique open palm, the beats in lines 28–29 shift to downward pointing index finger beats with the right hand. Why does Black reconfigure beat structure from open palm vertical to downward index finger

---

16 According to Lakoff and Johnson (1980) this reveals a "more of form as more of content" metaphor.

4.4 Poetic oratory as microcosmic ritual — 77

Figure 4.6: line 26 (high beat upstroke on *throughout*).

Figure 4.7: line 26 (beat downstroke on *treatment*).

beats? Why the shift from oblique open palms in lines 26–27 to index finger beats in lines 28–29 that accompany *when he died* and *funeral*? As Lakoff and Johnson (1980: 15) indicate, metaphors with an "up-down spatial orientation conceptualize bodies in the environment." "Up" is metaphoric of "health and life" while "down" is metaphoric of "sickness and death."[17] Yet there is no *a priori* reason to limit metaphoric conceptualization to language; the downward pointing beats represent an embodied metaphor for death.

---

17 In their classic text (Lakoff and Johnson 1980: 15), the downward spatial metaphors "his health is declining" and "he dropped dead" would be examples.

In lines 30–35 Black employs a rhythmic resolution to signal the end of the sacred performance: *Ted Kennedy went every chance he could get to be with the Smith family, because once again another widow in the family, another case of children without a father.* In line 33, he adds widows to the growing victim category; Jean Kennedy Smith takes her place alongside Jackie and Ethel as Kennedy widows – but with a rather novel twist. Typically a family has one widow but *another widow in the family* (line 33) refers to an extended kinship network rather than a simple nuclear family. Black does not make sole reference to the Senator's sister or the recently widowed Jean Kennedy Smith (or William Smith and his siblings) but to *another widow, another case of children* . . . Both are worked into the "another" categories, linking present to past and foregrounding widows and fatherless children in the process (and making the deaths more tragic). Both paired segments are organized around the indefinite determiner (*another* plus noun) to not only link widows and their children to the extended Kennedy family but, more importantly, to the distribution of collective memory and cultural trauma as well. Put another way, Black's mobilizes poetically regimented collective memories in which clause extension functions iconically to expand the pool of victims and co-victims.

And those memories emerge through the integration of gesture and language, a cross-modal pattern unfolding as follows. First, Black shifts back to open palm vertical beats in line 30 after the index finger beats in lines 28–29. Second, both paired segments (line 33 and 35) occur with phrase final pausing (1.4 seconds in line 33 and 3.5 seconds in line 35). Third, and most impressively, the paired segment beats in lines 33 and 35 consist of lateral movements (with the palm down or open hand prone engineered through the *flip-flop* or oscillating gesture that accompanies *once again* in line 32) rather than the orthodox up-down gestural beats in lines 30–31, which are segmented into equi-measured lateral increments that we saw previously in the listing items from lines 15–20 (moving left to right in 33, right to left in 34, further to the left and then to the right and further right in line 35). As this happens, the *traversing* beats mark distance to accumulate widows and fatherless children, visualizing the increase in number (see Figures 4.8 and 4.9). Here, the paired segments (*another widow, another case*) show (once again) how beat gestures incorporate more than rhythmic foregrounding and convey metaphoric images of accumulating numbers in a multidimensional and polyrhythmic poetic format (the lateral pattern adds numbers). More sharply put, the rhythmic pulse becomes more insistent, more prominent through the indefinite determiner by linking the assassinations of J.F.K. and R.F.K. to the cumulative widows and fatherless children.[18]

---

[18] While twenty-one children are without a father and several widows are without husbands, millions of Americans are without their cherished leaders – a psychic wound that shaped a generation.

The *once again* adverbial along with its gestural counterpart warrants a more technical appreciation than provided thus far. In lines 30–31 Black notes that *Ted Kennedy went every chance he could to be with the Smith family*. He then expands the clause with the *because* subordinate and temporal adverbial *once again*. However, the *reason*-clause is more or less gratuitous since he has already mentioned that the Senator frequently visits his sister since the death of her husband. Why expand the clause to provide the reasons in lines 33–35? Consider the temporally patterned gestures and interpretation of time-space, especially how the oscillating gesture that co-occurs with *once again* encodes the horizontal construction of time: the future to the right of the body, past to the left, and present to the center (Cooperrider and Nunez 2009).[19] In this respect, Black launches a temporal sequence marked in both speech and gesture where, first, the *once again* adverbial denotes repetition of some action or event and, second, wrist rotation from open palm vertical to open palm prone and then back to vertical toggles temporal alignment between current speech, past recapitulation, and future speech – speech designed with an eye toward the future outcome of the trial. In essence, the integration of the *reason*-clause and gesture functions as an ingenious method to *naturalize* a collective representation of the Kennedy assassinations without being too conspicuously strategic. For as Komter (2000: 420) mentions: "The task of both the prosecution and defense is to present the jury with the more convincing story. The problem is that too conspicuous an orientation to 'winning the case' might undermine the persuasiveness of their story." In more theoretical terms, Black employs a denotationally implicit (rather than explicit) form of cross-modal interdiscursivity that makes his argument persuasive. By mapping this "stealth" interdiscursivity onto legal argument, he smuggles in an emotional representation of political assassination indirectly under auspices of merely stating that the Senator goes to visit his bereaved sister and her children.

4.5 Section summary

In this section we have seen how multimodal action invokes the macroscopic order. If one site for the production and reproduction of collective memory and cultural trauma lies in ritual performance, and verbal and gestural cross-modal parallelism represents the "hallmark of ritual textuality," (Fleming and Lempert

---

[19] According to Cooperrider and Nunez (2009: 184), speakers encode time through spatial-directional gestures in which "time runs from left to right."

Figure 4.8: line 33 (low arc lateral beat on *another*).

Figure 4.9: line 33 (low arc lateral beat on *widow*).

2014: 491; Stasch 2011; Travarez 2014), then Black's projection of sacred images linking present to past certainly warrants inclusion in what we conceptualize as group memory. We have seen thus far how multimodal resources and dense polyrhythmic (and multidirectional) patterns are brought to bear on the construction of such memories, more explicitly, how the form of multimodal discourse itself – not the mere substance or topic of that discourse – participates in the production and reproduction of the sacredness of the occasion as it enlivens and intensifies collective sentiments. And that sociocultural function possesses – like all ritual performance – transformative consequences. The trauma experienced by the victim of rape is expanded and transformed into the collective trauma of cohort-driven experience, including in this instance the jurors in the trial.[20] It expands the pool of victims to include not just Bowman but the defendant, the Kennedy family, and the nation at large, as it transforms the focus of the legal case from the rape victim to the Kennedy family tragedies and the cultural trauma experienced by the collectivity.

4.6 The interactive co-construction of collective memory and cultural trauma

In the previous section we examined how collective memory and the cultural trauma embedded in it inheres in the narrative of a single speaker during opening argument. In this section, we show how sacred images and collective sentiments emerge as an interactive co-construction between attorneys and witness in direct and cross-examination. Consider the following examples.

Example 2 Ted Kennedy (TK) Direct Examination at 2:25 by Prosecuting Attorney (PA)

```
01    PA:    What led to your going to Au Bar tha- (.) that particular evening.
02                    (0.6)
03    TK:    . . . I and my sister Jean (.) and uh::: Bill Barry and Mary Lou
04           Barry (1.3) spent a couple of hours together at uh:: (.9) at the
05           end of the patio . . . The conversation was very (0.5) emotional
06           conversation, a very difficult one (.8) brought back a lot of uh::
07           very (0.5) special memories to me (1.5) particularly with the loss of
08           Steve (.) who uh really was a (.) brother to me and other members of
09           the family (1.0) and(hhh) (1.6) I found at the end of that conversation
10           I was (1.2) not able to uh::: think about sleeping (2.0) It uh was a
11           very draining conversation (1.6) a whole (.) range of memories (1.4)
12           really were (.7) came as an overwhelming wa::::ve of in terms of
                                   [            ]
                                ((Sweeping Gesture w/right hand))
13           emotion . . . I needed to talk to Patrick or William . . . So we
14           went to Au Bar.
((Gesture in line 12 brings right hand to center of body and sweeps to the right laterally))
```

---

20 According to Corning and Schuman (2015: 102) the assassination of J.F.K. is the fourth most frequently remembered event in the "life of the individual" (after Vietnam, World War Two, and space exploration) for those born between 1933–1953 (precisely the age cohorts of jurors in the Kennedy rape trial). Eyerman (2012: 565) also reveals the shocking experience of JFK's assassination for Americans based on public opinion surveys.

Example 3 Ted Kennedy (TK) Cross-Examination at 39:20 by Defense Attorney (DA)

```
01   DA:   You mentioned that ta- the::: (0.4) uh:: (.) that Bill Barry
02         and his family (.) were staying (.) at the home with you over
03         (.) Easter weekend
                        (1.0)
04   TK:   Yes.
05   DA:   Could you tell us who (.) Bill Barry is and what his relation to
06         to the family is.
07                      (1.0)
08   TK:   Well Bill Barry was uh::: (.) former (.) FBI agent tuh who uh::: (.)
09         now specializes in (.) in security matters (1.1) uh He was
10         probably (.) Jean and Steve's one of their two or three best
11         friends. He was one of my brother Bob's (.) very best friend.
12         And (.) he had provided uh:: security for (0.8) uh:: my brother
13         Bob (1.2) in the uh Nineteen (0.5) Sixty-Eight campaign.
14         And (0.9) he was with *my brother* (0.8) when he was **killed?**
           ((stress and very high pitch on both stressed items in line 14))
           ((lower volume in 14 except on **killed**))
15                      (1.8)
16   DA:   Does have have uh (.) special relationship with the family=
17   TK:   =**Very very** special.
18                      (7.1)
19   DA:   In fact is he not the (1.7) uh::: man who knocked the gun out of
20         *Sirhan Sirhan's hand* ((much lower volume))
                        [
21   TK:               ((4 micro head nods overlapping turn final component
                        and into a 1.0 second pause before the answer below))
22   TK:   Yeah ((very low volume))
```

Example 4 TK Cross-Examination by Defense Attorney (DA)

```
01   DA:   You said that ta- that evening (2.3) you had a uh (2.7)
02         sort of an intense (2.2) conversation with Bill Barry and
03         your sister Jean Smith (0.8) what was that conversation
04         about.
05                      (2.6)
06   TK:   ((In breath/out breath)) Well I think I uh:: described
07         it tuh (1.5) earlier (1.3) uh (2.6) uh:: (11.1)
08         ((2 micro head nods)) *I think I described it earlier* ((whispered))
```

Example 2 begins with the prosecuting attorney (PA), Moira Lasch, asking Senator Kennedy why he went to the Au Bar night club on the evening in question. Her *wh-*question opens the door for an emotional response in the next turn. After a short pause, he indicates that he went to Au Bar because of an *emotional conver-*

sation (lines 05–06) that left him troubled and anxious, needing someone to talk to about the unsettling memories that emerged during the evening. Although the Senator never elaborates the specific contents of the conversation, he does mention the loss of his brother-in-law, Steve Smith, (among other items) leaving open the possibility that the *special memories* might index the loss of other family members. At the very least the *loss of the "brother" Steve* links the loss to other brothers in the Kennedy family – a distinct possibility since, as Eyerman (2011: 71) mentions, Steve Smith along with Ted Kennedy organized R.F.K.'s funeral.

In lines 11–13, the Senator employs two metaphors that led to the decision to go to Au Bar, the first a *draining conversation,* the second *an overwhelming wave in terms of emotion,* once again without specifying the contents of the conversation. Both metaphors provide allusive reference to significant memories of family tragedies.

As mentioned previously, metaphor refers to the figurative aspect of language, and how an abstract or target domain, such as emotion, maps onto a more concrete or source domain. First, and with these points in mind, *conversation* (in line 11) is like a container with the content – in this case intense emotions – draining out, much like liquid draining from a container. Second, *emotional memories* (lines 12–13) *overwhelmed* the Senator like being swept away by powerful sea waves, an uncontrollable force of nature engulfing whatever lies in its path. As a consequence, he was unable to sleep and needed to talk to his son or nephew (line 13). Such intense emotions index allusive images of the personal and collective traumas that stimulated them, creating a heightened sense of emotional involvement, drawing the audience – the jury – into the emotional experience itself.

But as we saw previously gesture and other modal resources may integrate with language to convey powerful metaphoric images. In line 12, the Senator recruits a "sweeping" gesture with the right hand (positioned in the center front of the body and moves laterally a couple of feet to the right) that accompanies *overwhelming wave* (see Figure 4.10 below). Moreover, *wa::::ve* consists of vowel lengthening to further visualize *emotion* in motion (covering distance), demonstrating how language, gesture, and prosody intersect in the production of metaphoric imagery. That is, the utterance enacts the wave in motion, showing the analytic value of looking at communicative practice with all multimodal cylinders firing simultaneously in the production of meaning. Thus, metaphors occur not only in language but also as embodied action in and through the integration of language, prosody, and gesture to link personal grief with collective trauma, metaphors designed to *overwhelm* jurors with the same *emotional wave* that swept over the Senator. In the case here, the only force that powerful and overwhelming would be emotional memories of an acutely intense nature.

Figure 4.10: Example 2, line 12 (hand gesture on *overwhelming wave*).

Turning to example 3, Black resurrects testimony from direct exam in his cross. In lines 05–06, he asks about the identity of Bill Barry and *what his relation to the family is*. While the Senator's response to Lasch in direct exam was more or less restrained and composed, his demeanor changes quite dramatically when Black asks about Bill Barry.[21] In line 11, the Senator mentions that Barry was *one of my brother Bob's very best friends*. In so doing, he encodes a role relational reference with the possessive determiner, kin title, and diminutive (*my brother Bob*) that creates an emotional stance in a way that R.F.K. or Robert or any other reference would not. More explicitly, the diminutive marked address conveys a sense of intimacy and closeness, indexing the personal with the collective and, more speculatively, creates an inferential link with the other brothers, such as J.F.K.[22]

The Senator repeats the possessive determiner plus kin title in line 14 just prior to the clause final relative (*when he was killed*). However, this time the possessive and relative shift to much higher pitch and tremulous voicing to index a striking degree of emotional intensity and grief (see Hepburn and Potter 2012: 198–200). The Senator's response appears as a "holding back tears" voice quality that integrates with facial expressions of emotional pain: the inner corners of the eyebrows raised, eyelids lower, lip corners pulled down, furrows across the brow, and wrin-

---

[21] Bill Barry was RFK's bodyguard during the 1968 Presidential campaign and got separated from him while helping the pregnant Ethel Kennedy off the podium after the victory speech that fateful evening of June 5, 1968. Barry never forgave himself for not taking the bullets meant for RFK.
[22] R.F.K. and J.F.K. lived parallel lives in many respects, making it difficult to speak of one without the other. They are even buried next to each other in Arlington National Cemetery.

kles across the forehead (see Figure 4.11).[23] In line 16, Black asks if Barry *has a special relationship with the family*, and the Senator's immediately latched response double-ups on the adverb intensifier and employs a creaky voice quality – v*ery very special* – to foster a solemn emotional atmosphere to the proceedings.

Figure 4.11: Example 3 line 14 (sadness expression on *killed*).

However, the solemnness of the occasion is not the product of a sole speaker's contribution but a thoroughly interactive co-production: an intuitive duet between the attorney and witness that adds a tranquil spell to the proceedings. In lines 19–20, Black shifts to a much lower volume over the clause final possessive (*Sirhan Sirhan's hand*). In response, the Senator employs four micro head nods over Black's turn final component that continues into the 1-second pause before his barely audible response (*Yeah* – see Figure 4.12). Just as crucial, on the affirmative token the Senator looks down and squints his eyes in a distant gaze display held in place for several seconds, conveying a further sense of embodied grief in the process. That both attorney and witness lower speech volume at key moments indexes the interplay of microcosmic action and macrocosmic order in a grief-filled ritual of lament.

In the final example, Black attempts to unpack the Senator's gloss from the PA's question we examined previously: the content of the *draining conversation*. In lines 01–04, his question consists of delays, pauses, and hedge to index the delicate nature of the conversation. More explicitly, the question occurs with pauses ranging from 0.8 to 2.7 seconds, the *sort of* hedge or degree adverb, cut-off repair

---

23 See Ekman and Friesen (2003), especially chapter 9 and Ekman (2003) chapter 5 for in-depth discussions on the display of sadness.

Figure 4.12: Example 3 line 22 (sadness expression on *yeah*).

and empty filler *uh*. All these conversational devices display a sense of reluctance by delaying question production.

By the same token, the Senator's reluctance marked response aligns with Black's question form in a saddened sentiment that radiates through the contributions of both participants. His lip corners pulled down, lowered eyelids, and wrinkles across the forehead (with furrows across the brow also) index sadness, while the turn onset in and out breaths, stance verb (*think*), lengthy pauses/delays (including *Well*), thinking face gaze (over the 11.1 second pause), and whispered response (line 08) reveal the Senator trying to hold back tears as he speaks (see Figure 4.13) (see Hepburn and Potter 2012: 199). The emotional intensity is of such *overwhelming* magnitude that the grief-stricken Senator cannot even articulate the contents of the draining conversation. After closing his eyes as if too traumatized to respond (line 07), he merely repeats *I think I described it earlier* in line 08 from lines 06–07 (keeping in mind that he never did specify the content *earlier* but merely provided metapragmatic metaphors as we have seen – see Figure 4.14).

Yet, even as he conceals the content – never unpacking the metaphoric gloss – he reveals the intensity of emotional trauma, a socio-legal strategy that serves to stimulate the jury's understanding of its collective significance. There can only be one thing so emotionally intense as to be unspeakable: the assassinations of R.F.K. and J.F.K.

In fact, the current question/answer sequence appears *overwhelmed* by the same *wave of emotion* from the historical conversation he spoke of on March 30, 1991 (example 2). In this interdiscursive scenario, the current conversation maps denotational text onto interactional function, such that what we are doing *now* represents what was occurring historically (Silverstein 1998). That is to say, his response (re-) enacts the emotional trauma from the historical conversation that

4.6 The interactive co-construction of collective memory and cultural trauma — 87

Figure 4.13: Example 4 line 7 (*empty stare gaze away from Black in 11.1 second pause*).

Figure 4.14: Example 4 line 8 (closed eyes just prior to *I think*).

he is representing in the current here-and-now questioning: a powerful affective and epistemic stance that allows jurors to step into historical tragedy with all its emotional intensity as if happening in the current moment.

One final observation warrants consideration. The solemn ceremony involves co-operative action among Black, Lasch, Judge, and the Senator to manage the integrity of the sacred boundary. That sacred occasion is confirmed by both attorneys and judge who not only withhold interruptions but also, and most importantly, any objections during the silence (as in example 4, line 07). Indeed, both attorneys could object to witness testimony for failing to answer the question (and the judge could demand an answer or reply instead of the mere response). Nor do they "prod" the Senator to unpack the details of the *emotional* conversation (lines

05–06 in example 2), as they "honor" the gloss in an interactive confirmation of the sacredness of the occasion. In the process, they signal their participation in the solemn ritual commemorating the Kennedy tragedies.

4.7 Summary

One of the greatest speeches in American history was delivered on the evening of April 4 1968 in the heart of the inner city in Indianapolis, Indiana. In a thoroughly improvised speech, not written out beforehand, Robert F. Kennedy announced to the African American audience that Martin Luther King had been assassinated a short time earlier that night.

> "For those of you who are black and are tempted to be filled with hatred and mistrust of the injustice of such an act, against all white people, I would only say that I can also feel in my own heart the same kind of feeling. I had a member of my family killed, but he was killed by a white man."

Although he never spoke of his brother's death in public prior to this occasion, R.F.K. used J.F.K.'s assassination to form a collective, co-victim identity with those in the Indianapolis audience. As it turned out, while all the major cities in the U.S. experienced riots, death, and destruction, Indianapolis remained calm, and some claim it was because of the above speech. Only a couple months later, Ted Kennedy would deliver the eulogy for R.F.K. after he was assassinated. In a line just as poetic as Black's opening oratory, he delivered a moving image of collective lament for his brother, a lament strikingly reminiscent of example 4 when he speaks in the present moment about the assassination.[24] With his voice cracking with emotion and choking back tears he stated that his brother should be remembered as a:

> good and decent man who saw wrong and tried to right it, who saw suffering and tried to heal it, who saw war and tried to stop it.[25]

Black's questions (example 3, lines 06 and 16) and Senator Kennedy's testimony (example 2, line 09) use the same co-victim strategy but here with an eye toward

---

[24] Here we see the multimodal and affective interplay between form and substantive topic in the production of collective memories pertaining to co-victim identity.

[25] Senator Kennedy's eulogy at his brother's funeral was just as poetically eloquent as Black's opening remarks. Using a multi-layered poetic structure he produces listing repetition with embedded contrasts: repetition of the *who*-relatives, perception and effort verbs with the post-predicate infinitive clauses. Each of the effort verbs occurs stressed with increased loudness prior to the anaphoric pronouns (*it*), creating another dimension of opposition. In a final rhythmic drive, he creates metaphoric oppositions with *wrong/right*, *suffering/healing*, and *war/stop war*.

the legal-evidential work it may accomplish. More specifically, their choice of a particular kinship term refers to neither a nuclear family nor even an extended kin network like the "Kennedy family" but to *the family*. Why use this particular reference? Why this choice for kin classification? *The family* invokes strategic reference to "America's (Royal) family" Eyerman 2011: 70–71) a classification that includes not only the Senator but also the defendant and jurors as collective co-victims – co-victims in *the family's* tragedies.[26] In this instance, collective orientation functions strategically to shape testimony, evidence, and perhaps even the ultimate issue of the case in an emotionally stirring lament. It erases the defendant's agentive role and recalibrates socio-legal identity into the co-victim category. It shows how the adversary trial is not just a fact finding engine but, at times, a solemn ritual of collective memory and cultural trauma with historical voices positioned in the here and now of interdiscursive performance.

---

[26] As mentioned at the outset, since the trial was televised nationally and globally, the ritual of collective grief allowed millions of viewers the opportunity to participate in the sacred ceremony, not just those present at the trial.

5 Language, gesture and power in closing argument

In chapters 3 and 4 we studied the organizational dynamics of opening statement. In chapters 5 and 6 we turn to closing argument. According to both forensic linguists and legal scholars, closing constitutes the most dramatic moment of the adversary trial. Rosulek (2010: 218) refers to closing as the "master narrative of the crime, investigation and trial." Stygall (2012: 380) characterizes it as a "powerful moment" in the proceedings. Heffer (2010: 212) states that it is "often considered by trial lawyers as their main performance event . . ." Similarly, legal scholars emphasize the importance of closing argument for evaluating testimony. Mauet (2010: 387) mentions how closing represents the "culmination of the trial": an epistemological crescendo where jurors "are looking to see which lawyer *really believes* his side should win" (Mauet 2010: 398); and Tanford (1980: 133) describes how closing allows attorneys "to organize and emphasize favorable evidence, rebut your opponent's allegations, suggest ways the jury can resolve conflicting testimony, explain the law and show how the testimony leads to a verdict in your favor." Perhaps the main function of closing is to weave disparate strands of testimony from examination into a persuasive and coherent narrative.

Although closing represents the legal crescendo of a case, researchers like Rosulek (2008: 549) find that it remains "understudied" by comparison with other discursive events in the trial. Just as ironic, the few studies that focus on closing argument limit their analysis to speech, despite Mauet's (2010: 394–400) recommendation that attorneys should use "forceful" and "persuasive" gestures "to highlight major points of argument."

In this chapter we examine how gestures modulate the affective intensity and epistemic certainty of speech. We analyze how gestures synchronize with speech not only to orchestrate the rhythm of utterances but also to persuade jurors of the truth of one side and falsity of the other. We demonstrate how closing functions as a multimodal narrative in which gesture, gaze and speech transform legal evidence into institutionally organized forms of persuasive oratory. Reminiscent of Bulwer's *dialect of the fingers*, we show how attorneys foster the impression that they really believe their side should win by organizing and emphasizing favorable points of evidence, more explicitly by counting off and counting up inconsistencies in the prosecution case.

5.1 The relevance of gesture and multimodal conduct for closing

We have seen that a mere glimpse of trial interaction reveals that attorneys and witnesses often beat out the rhythm of their words to dramatize evidential points of significance, bestow an affective or emotive stance to their words and, as we will see, cumulate and quantify key strands of testimony. Legal actors may use gestures in concert with both gaze and speech to point out focal objects of attention. In fact, it is rare to find a moment when courtroom participants fail to gesture and/or deploy other forms of multimodal conduct in concert with speech.[27]

5.1.1 Beat gestures

Given its crucial role in the organization of communicative practice, how does gesture function in concert with speech? Once again, McNeill's typology of "iconic, metaphoric, deictic, beat quartet" provides an analytic point of departure (McNeill 1992, 2005). Iconic gestures resemble their referents; metaphorics depict abstract content; deictics locate referents through pointing; and beats orchestrate the visual rhythms of speech and foreground important information. This last gestural form is our focus in relation to closing argument.

McNeill (2012: 15) defines beat gestures as "yellow highlighters that beat out musical time" and in the process stress significant syllables and words. Like an orchestra conductor's baton, they parse speech into prominent, visual segments (Streeck 2008). As we have seen in previous chapters, experimental research confirms that beats not only signal emphasis but also influence the production and interpretation of prosodic prominence in their speech counterparts.

In this chapter, we show how a novel type of speech-synchronized beat not only orchestrates the rhythmic pulse of closing argument and foregrounds key strands of evidence but also, more substantively, cumulates and expands inconsistencies against the prosecution's case. That is, such beats engineer emergent forms of propositional imagery in addition to and simultaneously with their more orthodox rhythmic function, producing a multifunctional dimension of meaning that cannot be conveyed as richly through speech alone.[28] Along a different yet

---

[27] Of course in the U.S. witnesses must raise their right hand to take the oath before they can testify.
[28] To reemphasize our points on beats, according to McNeill (2005: 40) beats exhibit little structural variation in form and consist of simple up-down "flicks of the hand(s)." While they mark points of speaker emphasis they do not represent or depict semantic content. In other words, they are non-

simultaneous multimodal dimension, the attorney superimposes gaze-shifting movements of participation onto rhythmic beats to foster the impression that he *really believes his or her side should win*. Section one of the chapter demonstrates how precise coordination of speech with gesture projects multiple inconsistencies in the opposing side's case: how co-speech beats contribute to the epistemic calibration of certainty and hyper-persuasive affect through the poetic delivery of the unfolding text. Section two elaborates the use of this rhythmically infused gesture and shows how it functions ideologically as well as persuasively in oratorical practice. As the defense attorney lets his fingers do the talking he not only magnifies the prosecution's inconsistencies but also shapes and quantifies them in poetically organized and embodied forms of discursive power. Both sections expand on our investigations of beats thus far in the book and explore their sophisticated organization in much greater detail.

5.2 Background themes

In closing argument Black raised two themes that we address in this chapter. First, the prosecuting attorney claimed that Smith had deliberately evaded police requests for submission of evidence. Second, although the rather large and athletic Smith tackled the petite Bowman on the lawn of the estate, there was neither damage to her clothing nor marks on her body from the collision. As we will see next, Black uses these twin themes to develop inconsistencies in the prosecution's case through a novel form of staccato-like gesture that hammers home a significant piece of evidence; what we have referred to as *interdigital beats* or the right index finger moving up and down the digits of the left hand.

5.3 Interdigital beats and obstructing justice

We begin with example 1 below, in which Black argues that, in contrast to the prosecuting attorney's closing argument, the defendant never obstructed the police investigation, even surrendering voluntarily when charges were brought.

---

imagistic (Kendon 2004: 99-100). As we (Matoesian & Gilbert 2015) have shown previously, however, and as we show here, beats are much more complex than McNeill and others such as Goldin-Meadow (2003:8) have indicated. Rather than functioning as off-propositional flicks of the hand, beats may, at specific moments, exhibit considerable structural variation in form and invoke semantic imagery along with their orthodox rhythmic function and may do so simultaneously. In our view, beats are, at specific moments of discourse, multifunctional.

Example 1

```
01   And what happens when he finds out about the allegations?
              [((two open palms in front of body))
02   He calls (0.5) and talks to people at the home
              [((left arm lateral open palm beat and hold)
03                        (.)
04   He gets a lawyer (.)            who immediately gets in touch with them
              [((left arm lateral open palm beat))    [((lateral open palm beat))
05                        (.)
06   He voluntarily
              [((hits left little finger with right index finger and holds))
              [((gaze at little finger for 0.5 seconds over the volitional adverb))
07   turns over all the
              [((gaze returns to jury from the gaze at the finger))
08   samples that they want
              [((IB=interdigital beat; pushes down little finger with right index finger))
09   photographs
              [((IB on ring finger))
10                        (0.6)
11   uh::: (0.4) everything.
              [((IB on middle finger and hold))
              [((highest elevation on upstroke over delay marker+vowel stretch))
```

In line 01, Black poses a rhetorical question synchronized with a two-handed open palm gesture (an open palm supine, Kendon 2004). According to Muller (2006: 244), such a gesture possesses an emphatic function, offered here as a puzzle that the defense attorney proceeds to answer with a poetic listing of items (see Figure 5.1 line 01 below). In lines 02, 04 and 07 Black crafts a poetic metricalization of third-person pronouns and action verbs that generate a rhythmic listing of items (*he calls, he gets, he turns over*), highlighting the significance of each individual piece of evidence in the series. Expansion of the listing tokens (or particulars related to the general interrogative *what happens* in line 01) projects onto the extralinguistic world of evidence to produce a cumulative effect against the prosecution's claims about obstructing justice: that the defendant never hindered the police investigation but cooperated voluntarily. Moreover, the attorney employs the phrasal verb in line 07 (*turns over*) as the interactional platform to launch an embedded series of noun phrases in lines 08, 09 and 11 (*the samples, photographs, everything*), with noticeably marked stress on each for emphasis.

Figure 5.1: line 01 (*happens*).

In lines 02 and 04 he inserts an even more intricate lamination, or embedding of poetic detail, consisting of multimodal layers of inclusion. The defense attorney repeats a lateral open palm beat with the left arm fully extended and synchronized with the respective lexical affiliates: *calls, gets* and *gets* (see Figure 5.2 line 02 below). The upstroke of each beat occurs on the pronoun in lines 02 and 04 and the *wh*-relative in line 04 while the stroke or meaning bearing phase of the gesture occurs on the action verb.

Figure 5.2: line 02 (*calls*).

Moreover, the phrasal motion verb in line 07 (*turns over*) not only serves as the staging point for launching the laminated pattern of NPs but also acts as a rhythmic anchor delivering an ascending progression of gestural beats that unfurls incrementally from the loosely furled hand, a process that unfolds as follows.

In line 06, Black activates space-shifting movements from the extended left arm to a repetitive series of *interdigital beats*, in which the tip of the extended right index finger lands on and ascends up the unfurling fingers of the left hand. To contextualize the shift, he coordinates beat on the little finger with marked gaze over the volitional adverb (relating to the willingness of the agent) and little finger (see

Figure 5.3 line 06 below). As we have seen from Muller (2008: 236): "Directing the gaze at something . . . indicates speaker's focal attention" and "turns the gesture into an interactively significant object." In this instance, gazing at the little finger while simultaneously hitting it functions as a pointing gesture to signal that the gesture possesses significant evidential import and the jury should also look at it (see Streeck 1993: 286). Moreover, he maintains gestural position with the right hand index finger latched onto the little finger of the left hand as he shifts gaze from the finger to the jury on the phrasal verb in line 07.

Figure 5.3: line 06 (*voluntarily*).

However, the significant object of attention is not merely the index finger hitting the little finger of the opposite hand. More importantly, the attorney's gaze draws attention to the forthcoming interdigital list superimposed over the NPs (*samples, photographs, everything*), which the beat on line 06 activates with the hit gesture.

In line 08 Black delivers an ascending progression of interdigital increments, which begins (as noted above) with the tip of the right index finger pushing down the tip of the little finger on the left hand. As we have seen, the beat lands on the volitional adverb, well ahead of the NP token (*samples*). As a demonstration of the temporal synchrony between speech and gesture Black "holds" the gesture – with the right index finger merely touching the little finger of the left hand – to manage the online pace of the gesture-speech ensemble, adjusting and readjusting timing in order to synchronize precise coordination with utterance construction. When the noun finally "arrives," the gestural hold on the little finger gets pushed down from the "landing" or touching position to coordinate arrival with its lexical affiliate. In so doing, he evokes a type of visual imagery that, when one object hits another, there will be a conspicuous reaction (a point we return to later).

The next beat ascends up the ladder of the unfurling hand on *photographs*, so that the right index finger hits and pushes down the ring finger of the left hand (see Figure 5.4 line 09 below). In line 11, the set-marking tag *everything* co-occurs with the beat on the middle finger and fosters an impression that other as yet unmentioned inconsistencies might be included under its auspices (see Figure 5.5 line 11 below). It is also synchronized improvisationally with its gestural counterpart, as Black makes online adjustments and readjustments in the speech-gesture ensemble to manage the interactional contingencies that arise (Kendon 2004). Similar to the beat on *samples*, the right index finger begins upstroke on the delay marker and lands on the middle finger during the .4-second pause. However, the tag has not arrived, so the index finger holds position on the middle finger till it does. Notice in particular how the delay marker and pause work as interactional resources to manage coordination between speech and gesture, making online adjustments to execute temporal spacing between the two modalities. That is, the right index finger maintains beat position to wait for the set marking tag to catch up, as if maintaining a holding pattern for one another, then hits and pushes down the middle finger. This demonstrates in vivid detail how speech and gesture are executed conjointly as a speech-gesture ensemble.

Figure 5.4: line 09 (*photographs*).   **Figure 5.5:** line 11 (*everything*).

In this rhythmic interplay of fingers, Black provides evidence that the defendant, unlike stereotypic beliefs about rapists, cooperated voluntarily with the police investigation. In the visual infrastructure where such sociocultural identities emerge, he orchestrates a polyrhythmic and multimodal expansion of inconsistent evidence against the prosecution's case on the issue of obstructing justice. By 'polyrhythmic' here we mean the repetition of action verbs, NPs, beat gestures and marked stress on significant points of evidence, creating a state of involvement with the jury and sense of cohesion in the closing narrative. By 'multimodal' we mean the improvisational integration among gesture, gaze and speech that modulates the affective intensity and epistemic certainty of the attorney's persuasive performance. Together, these features yield an intricately balanced metricalization of oratorical power that not only magnifies the inconsistencies but counts them as well. Indeed, Black's unfolding segmentation of individual items allows jurors to focus attention on the discrete accumulation of favorable evidence contained in the list. Rhythmic repetition in the speech-gesture ensemble produces pragmatic emphasis in which each individual item of evidential significance stands out like each individual digit in the unfurling hand, allowing the jury to visualize the adding-up effect in a dynamic spatio-temporal movement that cannot be captured as richly through speech alone.[29]

5.4 Interdigital beats and the embodied resistance ideology

In her classic study of rape trial discourse, Ehrlich (2001) analyzed how defense attorneys organized their side of the case around what she referred to as the "utmost resistance" ideology: a powerful sociocultural practice designed to generate inconsistencies in the victim's narrative. Utmost resistance – a once-codified legal hurdle for the allegation of sexual assault – required the victim to physically fight off the attack until all resistance was overcome by overwhelming male force. By way of analogy, this section examines how the defense attorney mobilizes more powerful inconsistencies through interdigital beats in the production of what we call an *embodied resistance ideology*, a requirement that evidence of resistance must be visible on the victim's clothing.

As mentioned earlier, Bowman claimed that the large and athletic Smith (six foot three and two hundred pounds) tackled her on the lawn of the estate after chasing her at "full tilt" yet there was neither damage to her clothing nor marks on her body after (what Black referred to as) this "tremendous impact." In the examples below he raises the condition of Bowman's clothing after the impact.

Example 2

B^=Intradigital Beat or up down push on the same digit with right index finger latched onto left little finger and closed into a fist shape
IB=interdigital beat
01   The   examination of the dress was done minutely (.) by
       [((gaze at finger))   [((B^   [((B^ [((B^   [((B^ [((B^
                               [((gaze returns to jury
02   Barbara Carabello ((lowered volume))
              (0.6)

---

[29] In addition to Jakobson (1960: 356) and Silverstein, Tannen (1989: 50) finds that repetition is "evaluative and emphasizes a point . . . Repetition of phrases establishes a list-like rhythm, giving the impression that . . . it constitutes a long list, longer even than the one given . . . The evaluative effect of the list is to communicate that the speaker finds the length of the list impressive – and so should the listener." Here we can see that the poetic function applies to multimodal conduct as well.

03   There's no *grass* strains (0.4)
               [((IB little finger))
04   There's no *dirt* (0.4)
               [((IB ring finger))
05   There's no *particles* (.) of grass (0.3)
               [((IB middle finger))
06   There's no *mud* (0.4)
               [((IB index finger))
07   There's no *chips* of concrete (0.4)
               [((IB thumb))
08   no little   stones (0.3)
               [((B2=vertical parallel two-hand gesture))
09   *no **nothing*** (1.0) on this dress at all (0.5)
      [(((criss-crossing lateral gesture in front of body with both hands/palms down
      [(((eyes close/gaze downward on the double-negative))
10   N'yet this dress (.) ha::d (.) a *tremendous   impact*
         [((B2))   [((B2))   [((B2))   [((B2))   [((B2))
         B2=two hand parallel vertical (up/down) beats

Example 3

B1=One hand vertical/up-down beats/chops
01   They looked at everything with a ***microscope*** (1.2) n found absolutely nothing
                                    [((B1+hold))   [((B1))
                        (2.1)
02   Not a single   *grass stain* (.)
      [(((gaze at finger))   [(((IB starting with little finger))
03   no *abrasion* (.)
         [((IB
04   no *cuts* (.)
         [((IB
05   no *rip* (.)
         [((IB
06   no *mud* (.)
         [((IB on thumb))
07   no *dirt* (.)
         [((IB on index finger on down slope))
08   no *soil*
         [((IB on middle finger on down slope with 1.4 second post stroke hold))
         [(((1.4 second post stroke hold) on *soil*))
09   *Nothing.*
      [(((B2=two hands open palms vertical parallel gesture))

Example 4

B2=Two handed open palms vertical/up-down beats
```
01   In the  weave  of the  pantyhose (.) there's  ab         solutely
             [((B2   [((B2   [((B2                  [((B2     [((B2
02   no      evidence.
     [((B2   [((B2
                             (0.5)
03   There's no                        sand.
     [((gaze at little finger))        [((IB-little finger
                             (.)
04   There's no  dirt.
                 [((IB-ring finger
                             (.)
05   There's no  grass particles.
                 [((IB-middle finger
                             (0.5)
06   There's no  soil.
                 [((IB-index finger
                             (.)
07   There's no  mud.
                 [((IB-thumb
                             (.)
08   There's none whatsoever.
     [((B2=two hand vertical palms parallel))
```

In examples 2–4, the defense attorney uses a type (or general) category to start each narrative segment, then unpacks that type through token (or particular) details. In example 2, he organizes the type around a forensic expert's *minute* examination of Bowman's dress, while in example 3 he announces that the state forensic examiners even used a *microscope*. At a finer level of granularity (Example 3 line 01), Black leans his upper torso into the stressed type component *microscope* to deliver a litany of referential detail that explicates the type. The experts not only looked at *everything* but looked at everything *with a microscope,* finding *no grass stains, abrasions, cuts, rip, mud, dirt, soil* (lines 02–08). For still further emphasis, he produces a 1.2 second prolonged hold on the instrumental role (line 01, example 3), contextualized by an accelerated downward gesture and noticeably marked recoil off the bottom of the downstroke.[30] More formally, the hold gets built off the top of the recoil in the form of an open hand supine (or right hand palm up) gesture, as if to say: "what more could be done to find any damage to the clothing?" In this instance,

---

[30] Occasionally, the gesture stroke (or meaning-bearing phase of the gesture) will be followed by a post-stroke hold, which keeps the stroke phase in play, often for emphasis.

Black's gestural intensifier transforms the speech event (one of listing or describing) to an evaluation of the event being described (i.e. instructing the jury how to interpret subsequent speech). As Kendon (2004: 270) suggests, such a gesture "makes a comment on something that was just said" or in this case what will be said, producing an interpretative frame for the subsequent listing. Example 2 constitutes a similar case in point. Here Black executes a *minute* discursive analysis (lines 03–08) of the dress superimposed over the *minute examination by Barbara Carabello* (the forensic expert in line 1 who found *no grass stains, no dirt*, etc). Both cases demonstrate in vivid detail how referential text (or what we are saying about something) gets mapped onto interactional text (or what we are doing now) to deliver a persuasive evaluation with all multimodal cylinders firing (Silverstein 1998).[31]

Example 4, however, departs from the agentive role exemplified in examples 2 and 3. Instead it moves directly to material evidence: the victim's *pantyhose.* Even so, a striking similarity may be noted in type organization of examples 3 and 4. The defense attorney organizes evidential findings around the epistemic stance adverb *absolutely* (the experts *found absolutely nothing* and *there's absolutely no evidence*), signaling degree of commitment to and level of certainty about the proposition. But there is more to *absolutely* in both uses that bears scrutiny here, more than lexical/grammatical encoding of epistemic stance through the degree adverb. While *absolutely* signals epistemicity at the level of grammar, it also co-occurs, first, with two-handed parallel (palms facing each other) beats in Figure 5.7 and, second, with one-handed vertical chops in Figure 5.6 which further intensify Black's alignment to his words (see Figure 5.6 line 01; Figure 5.7 line 01 downstroke; and Figure 5.8 line 01 upstroke below). In example 4 (line 01), he produces three vertical (up-down) two-handed beats (palms facing each other with low elevation on the beat upstroke) on *weave, of* and *pantyhose.* After the contracted existential copula he repeats the same beat, though with an elevated upstroke – and increased acceleration on the downstroke – that lands on the first stressed syllable of the intensifier. Black repeats this marked gesture on the second syllable of *absolutely*, the *no*-determiner and noun in a dense flurry of two-handed vertical beats that convey emotional conviction. In example 3 (line 01) he uses (one-handed) right hand chopping gestures to accompany *with a microscope* and the sensory verb plus object (*found absolutely nothing*). Just as important, the highly symmetrical stress on the sentence-final nouns and indefinite in example 3 (lines 02–09), and on the sentence final nouns in example 4 (lines 03–07) adds a further layering of poetic detail to the developing inconsistencies.

---

31 Put another way, both instances display a parallelism between narrated and narrating events.

Figure 5.6: Example 3 line 01 (*microscope*).

Figure 5.7: Example 4 line 01 downstroke (*of the*).

Figure 5.8: Example 4 line 01 upstroke (*weave*).

After the type-indefinite introduction (*everything*), Black begins his listing process with marked gaze on the turn-initial first token (see Figure 5.9, example 3 line 02 below). As illustrated previously, the attorney directs gaze at the unfurling finger to start a polyrhythmic and multimodal pattern of evidential listing, gazing at the initial beat to coordinate a focus of joint attention with the jury. The deictic gaze that points to and locates attention in examples 3 and 4 aligns here with the same rhythmic pattern – on the existential plus contracted copula in example 4 (line 03) and negative particle in example 3 (line 02) – turning the gesture(s) into an interactively significant object. However, the gaze in example 2 begins on the turn-initial determiner of the type (line 01). Why does his gaze in examples 3 and 4 land on the token listing of items or the particulars of *everything*, while it co-occurs on the type-category in example 2? Why does example 2 depart in this way from the rhythmic pattern?

Figure 5.9: Example 3 line 02 (*grass stain*).

Or does it? The attorney's gestures in example 2 (line 01) involve not just any kind of gesture. They consist of the right index finger latched onto the little finger of the left hand (in a closed fist), and he beats out rhythm (over the words indicated) in this position (see Figure 5.10, example 2 line 01 below). That is, the type category interacts with a less marked, more subtle, series of fist-like beats, which is why the attorney directs gaze to the determiner (example 2 line 01). Instead of ascending to the ring finger the attorney maintains position with the right index finger latched onto the little finger of the left hand in an embedded lamination of delicate gestures – *intradigital beats* – that reflect and visually demonstrate the detailed and precise examination of the dress, highlighting *examination, of the dress*, and *done minutely* by landing on an explicit point of emphasis.[32] By using this gestural variation, he regulates the tempo of the interdigital beats for the forthcoming listing tokens (or particulars) to elaborate the type (or general category). Rather than deviating from the gaze distribution seen in the other examples, Black adheres to the pattern by gazing at the initial latching touch on the little finger, a rhythmic modulation of, rather than departure from, the gaze distribution.[33]

Figure 5.10: Example 2 line 01 (*examination*).

Thus, in each of the examples, the attorney simultaneously gazes at and hits the little finger of the left hand with the index finger of his right hand to contextualize the significance of the rhythmic pattern. He then shifts gaze to the jury and launches a polyrhythmic fusillade of tokens that elaborate or unpack the type by (1) repeat-

---

[32] To be more precise, intradigital beats occur with the index finger of the right hand latched onto the little finger of the left hand, beating out rhythmic significance over an entire idea unit with this latching motion in a closed fist position. That is, intradigital beats use a single digit or form for a single idea unit; they don't elaborate or unpack details. Detail-oriented interdigital beats, on the other hand, occur with the index finger of the right hand ascending up the ladder of the left hand (starting with the little finger) as each individual digit unfurls to emphasize the listing tokens. They use multiple digits for multiple idea units and unpack details visually.

[33] Furthermore, the linking of the little finger of the left hand by the index finger of the right hand prior to the onset of the token links the prosecution's examination to the findings – depicting Black as a mere relayer of information, effectively concealing his partisan role. We can see here how gaze and gesture combine to delineate participation roles in oratorical performance.

ing the negative + NP and (2) segmenting the digits into discrete units. Verbally, Black repeats the existential + contracted copula, negative and NP (examples 2 and 4 lines 03–07 *There's no dirt* etc) and the negative +NP (in example 3, lines 03–08 *no cuts* etc). Visually, he extends the fingers of the left hand (furled together in a fist-clenched position in the starting off position), such that each individual digit unfurls incrementally from the bottom little finger to the thumb (which is extended by default). Once unfurled, each digit is then hit by the extended index finger ofv the right hand, pushing it down in a pronounced movement of hitting impact (once again, a crucial point we will discuss later). Integrating both verbal and visual, his beat upstroke (in examples 2 and 4) co-occurs on the contracted existential copula while the stroke lands on the NP (see Figure 5.11, example 2 line 06; Figure 5.12, example 2 line 07; Figure 5.13, example 4 line 05 upstroke; and Figure 5.14, example 4 line 06 downstroke below). In example 3, the upstroke co-occurs on the negative while the stroke once again lands on the NP. Just as important, each interdigital beat co-occurs with stress in either its NP affiliate or the first modifier in the NP and ends with a short pause before moving on to the next item in the series, so that the tokens appear measured out in discretely layered intervals. Finally, as the index finger of the right hand beats up and down, each unfurling digit of the left hand extends outward from its coiled/furled position and stays extended after receiving its beat in a chopping position (looking like a karate chop that transforms the starting fist position).

Such interdigital beats function to individuate, expand and cumulate individual pieces of evidential significance through finger movements along the hand. In that process, counting may be viewed not as a static or given outcome of some mathematical operation but as an active micro-technique of oratorical power: counting as a situated and multimodal accomplishment. In all three examples, the manual modality adds another dimension to meaning that is not as vividly captured through the oral modality alone, merging epistemic with affective stance.

Figure 5.11: Example 2 line 06 (*mud*).   **Figure 5.12:** Example 2 line 07 (*chips*).

Figure 5.13: Example 4 line 05 upstroke (*particles*).

Figure 5.14: Example 4 line 06 downstroke (*mud*).

5.5 Type-token reflexivity

We have seen, then, that Black mobilizes multiplex laminations while integrating speech and gesture and builds tension in his narrative through amassed repetitions: a repetitive groove that starts with a type category as a way of launching the token items. There is also a form of embedding that merits more detailed discussion. In examples 2–4, Black produces not only type beginnings for each segment but type ending categories too, embedding tokens between the types. In this pattern, the type elaborates each unfolding token as each token elaborates the type, and provides an interpretive space for the next token in the itemized series. The segment-ending type resurrects and reaffirms the discretely anchored motifs in the type-token relationship.

In example 2, Black alternates the type-token pattern in line 08, from the turn-initial existential+negative (lines 03–07) to the turn-initial negative: a rhythmic reduction in lines 08–09 that maintains the sustaining texture while alternating rhythmic structure (*There's no chips, no little stones, no **nothing***). Moreover, he departs not only from the prior repetitive frame but also from the repeating inter-digital gestures (line 08) and employs a two handed vertical gesture with the palms facing one another that co-occurs with the adjective phrase, a shifting arrangement that injects a sense of novelty into an otherwise predictable pattern (see Figure 5.15, example 2 line 08 below). But what is most novel about his oratorical performance is something further. Black layers polyrhythmic modulations to produce the double negative with increased stress and volume in line 09 (*no **nothing***) that repeats the turn-initial negative from line 08, intensifying epistemic and affective stance about the forensic expert's analysis of Bowman's dress (from line 01).[34] In addition, he synchronizes the double negative with two multimodal components. First, he layers it with a two-handed crisscrossing lateral gesture with the palms down (see Figure 5.16, example 2 line 09 below). According to Kendon (2004: 225) this type of gesture serves a performative function: "Gestures with the open hand held so palm faces downward ... and is moved laterally in a decisive manner is an act of rejection or denial."[35] Second, he gazes down with his eyes closed to produce a multimodal display of affect (line 09), fostering an emotionally charged impression of truth, sincerity and spontaneity. In this multi-laminated maneuver, we see speech, gaze and gesture converge and merge to execute a powerful epistemic and affective stance.[36]

Figure 5.15: Example 2 line 08 (*stones*).　**Figure 5.16:** Example 2 line 09 (*no nothing*).

The double negative and crisscrossing gesture not only recalibrate epistemic and affective value; they also figure prominently in a tightly woven contrast in line 10. In a further layering of metric modulation, Black mobilizes the double negative as the first component in a juxtaposition of contradictory elements, whose second part employs the utterance-initial contrastive adverbial (*N'yet this dress had a tremendous impact*) as culmination of the argument: a rhythmic resolution synchronized with a dense flurry of vertical beats (two handed parallel palms facing each other in an up-down movement) to foreground the main points of the contrast, i.e. that given the *tremendous impact* one would expect evidential residue on Bowman's clothing (see Figure 5.17, example 2 line 10 below).

---

34 According to Biber et al (1999: 178), "Because of the repetition of the negative form this type of negative appears to have a strengthening effect. Here *no nothing* equals *not anything*." But of course it is just *not anything*. *Not anything* is a big something for the defense.
35 Harrison (2018: 39) refers to this as a "2-palms down across" gesture and not only refers to negative evaluation but to a situation when the speaker "wishes to finish an argument."
36 Just as interesting, Black uses the proximal demonstrative *this* in line 9 to bring the dress into a state of deictic immediacy for the jury.

Figure 5.17: Example 2 line 10 (*N'yet this dress*).

In examples 3 and 4 Black continues to elaborate the type-token relationship with a repetitive type-closing. In example 3, he produces a partial repeat (*nothing* in line 09, from the indefinite in line 01, *absolutely nothing*) which co-occurs with the two-handed open palm gesture, providing an "intensifier effect" on the stance-marked negative (Mueller 2004: 244). In example 4, he tweaks the type-beginning (from line 01) for his type-closing (in line 08), while maintaining continuity with the initial motif (*there's absolutely no evidence* and *There's none whatsoever*). Here the intensifier *whatsoever*, emphasizing the negative aspect of the statement, integrates with the two-handed vertical gesture to modify the quantifier *none*, adding a strong emphatic stance to the token items that serves to remove any ambiguity about the proposition.

In essence, it may be suggested that the type-token relationship seeks to establish its own sphere of truth and facticity. The type frames the tokens that are included while the tokens provide evidence for the type. Together the two levels yield a gestalt-like configuration which reflexively draws attention to the narrative through a densely laminated and polyrhythmic integration of speech, stance and gesture. Black's deft virtuosic twists and turns produce a dynamic that foregrounds affect by varying gestural form relative to the type-token relationship. That is, for type gestures he uses two-handed gestures; for the tokens he mobilizes interdigital beats to count off the inconsistencies or unpack the type.[37]

---

[37] We have noted that what might arguably be exceptions to this in the intradigital beats in line 01 and the gesture in line 08 in example 2. But perhaps we should temper such observations. The intradigital beats also express a single idea unit like the types (*The examination of the dress was done minutely by Barbara Carabello*). The two-handed beat on *little stones* appears designed to intensify the poetic repetition occurring on the double negative in line 09. One other consideration deserves mention here. The intradigital beats contain no stress over their lexical affiliates. On the other hand, the NP items that co-occur with the interdigital beats contain stress.

5.6 Multimodal modulations

As we have seen, Black orchestrates several multimodal modulations of the above pattern. First, in example 2 he recycles the intradigital beat into an interdigital beat to recalibrate and segment the listing process. After the hold on the address term in line 02 (*Barbara Carabello*), the right index finger lifts off the latching position and then moves down to hit the unfurling little finger on the left hand, a strategic maneuver that individuates the little finger from the former intradigital grouping.

Second, he rotates gaze from left to right in an incremental (horizontal/lateral) progression as if to engage each juror by coinciding eye movement with each beat stroke in lines 03–07 (on the thumb beat he recalibrates gaze direction one increment to the right).

Third, Black not only moves gaze direction (his head) in lateral increments, he also simultaneously incorporates head and upper torso vertically to build contrapuntal rhythms for heightened affect (beating with the head and torso up and down along with the first three interdigital beats in lines 03, 04 and 05): a multidimensional, multimodal and polyrhythmic layering of sound, imagery and movement.[38]

Example 3 reveals another variation, this time on the interdigital pattern. After the degree adverb and negative pronominal (*absolutely nothing*), the attorney launches an ellipted series of tokens that unpack what "nothing" consists of. Once again, he starts with noticeably marked gaze on the initial items in the adjective phrase (*not a single*) then redirects his gaze to the jury. In this symmetrical series he proceeds once again to unfurl his fingers in an ascending progression from the little finger up to the thumb on *mud* but then runs out of fingers as he scales the ladder of the hand! More prosaically, Black runs out of fingers on the upslope because there are so many inconsistencies. In this improvisational movement, he recycles the index finger and middle finger on the downslope to create an ascending and descending progression of interdigital increments.

Also germane to the above point, the 1.2 second prolonged post-stroke hold on the middle finger accentuates the significance of the number of inconsistencies. For McNeill the "hold ensures that the meaningful part of the gesture – the stroke – remains semantically active." Perhaps the hold here, however, does more than keep the meaningful part of the gesture in play. It emphasizes the quantity of

---

[38] We draw attention here to grammatical parallelism (along with stress and pausing), the right finger moving up and down, the digits of the left finger extended outward one at a time, lateral head movement and vertical or up-down motions of the head and upper torso that generate an intricately synchronized tapestry of sound, imagery and motion.

the inconsistencies in such a way that, although Black could continue, he pauses for all practical purposes as if to say: "This is getting ridiculous and even though we could continue we'll stop here because you get the message." Speakers can in this way encode and convey motion and direction – even stance in this context – more vividly through gesture than speech alone.

5.7 Residual semanticity

There is a further aspect of gesture not easily, or even possibly, encoded in speech. We have seen how the interdigital beats recruit the index finger of the right hand to hit and push down the digits of the left hand in the action-reaction imagery. We also mentioned at the beginning of the chapter that beats may, at certain moments in the discourse, not only foreground key strands of evidential significance but also evoke aspects of propositional imagery, i.e. of semantic content. It is time now to develop and conceptualize this multifunctionality of beats in more detail. According to the ideology of embodied resistance outlined above, expert scientists should have found forensic residue on Bowman's clothing because of the "tremendous impact" between her and the defendant. The fact that no evidential residue showed up under microscopic scrutiny exposes, according to the defense, a serious inconsistency in the prosecution's case. During the interdigital beats, the friction of two fingers colliding signifies the impact between one force hitting another and projects how there should be some visible consequence as a result. More explicitly, just as the interdigital beats represent a symbolic residue of touching (contact) so too should the collision between Smith and Bowman be revealed by evidential residue of the collision on her clothing. Put another way, Black shows how one force impacts another in the pushdown part of the stroke or stroke intensifier and how, by analogy, there should be residual effect on Bowman's clothing. In this action, the referential aspect of the text maps onto the here and now interactional-multimodal text in a dynamic process of what we refer to as *residual semanticity*; how what we are saying becomes what we are doing through the integration of text metricality and rhythmic beats. Beats, in this instance, not only serve a pragmatic function but convey residual semanticity or aspects of content as well.[39]

---

[39] Along similar lines, the ascending and descending movements on the hand (which show the increase in and direction of inconsistency) are not something easily conveyed in speech as in gesture.

5.8 Power and multimodal conduct in the law

In the adversary system, the weight of evidence refers to the weight or significance of facts necessary to tilt the burden of proof to one side or the other, and such proof rests on the persuasiveness of the evidence. Which attorney really believes his or her side should win? Who counts off the most facts and thereby creates the most favorable impression on the jury? And, most importantly, how does one attorney mobilize power to make their account *count*?

Consider counting in more literal terms. In examples 2–4, the token items Black counts off with the interdigital beats could suffice semantically at the type level (such as "There's nothing on her clothing" etc). Or he could convey the general meaning with some other gesture, such as a precision ring (in which the tip of the thumb touches the tip of the index finger on the same hand) or even a point (a vectorial indication in space). This demonstrates how the rhetorical force of the chosen gestures combined with other levels of language organization crystallizes and adds particularity to the evidential points – binding different levels of discourse together. When scrutinized in detail, the interdigital beats accompany synonyms or noun phrases similar if not identical in meaning at the conversational level: *grass stains* equals *dirt* equals *mud* equals *soil* and so on. What interactional work does such redundancy accomplish in the law?

The interdigital beats make each individual piece of evidence stand out, just like each individual digit stands out from the others. The beats carve up or dissect a unity into discretely layered and distinct items of significance: a microscopic division of referential detail that counts off and expands inconsistencies in the prosecution's case. That is, as the defense attorney lets his fingers do the talking he not only magnifies the prosecution's inconsistencies but also shapes and quantifies them in a poetically organized and embodied form of discursive power. In the spatio-temporal conversation (or in Bulmer's terms "Dialect of the fingers") among the fingers, he allows the jury to visualize that counting and cumulative process in motion.[40] This conversation demonstrates that counting is not merely a mathematical operation but a sociocultural resource in the construction of legal context.

More theoretically, we are claiming that counting, cumulating and differentiating are more vividly conveyed through gesture than speech. According to Goldin-Meadow (2015: 73): "Because the representational formats underlying gesture are mimetic and analog rather than discrete gesture permits speakers to represent

---

[40] Moreover, individuating the whole displays a level of precision that confers authority on his words and shows the jury that Black considers all these tokens important enough to mention individually, important enough to count off in gestural form.

ideas that lend themselves to these formats (e.g., shapes, sizes, spatial relationships) – ideas that . . . may not be easily encoded in speech." Just as important, touching the individual digits of one hand with the finger of the other hand encodes and individuates objects more effectively than other types of gesture. As Alabini et al find (2001: 52): "Touch is closer to the tagged object than a point; it is more clear which specific object is being indicated by a touch than a point." By touching each digit to branch off from the whole, Black signals that it warrants unique consideration as an individual item of evidence. He integrates repetition of the unfolding text with gestural rhythms to produce a semantic foregrounding affect (as mentioned above conveying substantive information and performing a pragmatic function simultaneously), a persuasive oratory that may well tilt the balance of proof and put the defense case in the most favorable light.

And that balance of proof poses important questions and problems in relation to the resistance ideology described earlier. Embodied resistance foregrounds inconsistencies in the victim's account in and through a polyrhythmic infrastructure of interdigital beats in concert with repetitive features of speech, making ideology of inconsistency invisible as a taken-for-granted process of power. In the law, inconsistency is taken as a natural incongruity or juxtaposition among contradictory facts of evidence. However, as Black unfolds the digits to individuate items of evidence from the whole (and thus increase their size/weight), he conceals their "factual" status as a unity of similar if not identical items, a multimodal ideology of language that naturalizes inconsistency in the mutual interplay of the fingers.[41] Black's beats impart a natural "feel" to his verbal rhythms and inject them – and the inconsistencies – with a powerful sense of oratorical precision. Just as crucial, inconsistency is framed from the defense attorney's perspective, while alternative possibilities remain unexplored or erased (e.g. that the victim was traumatized, that the defendant picked her up and put her on the ground or that perhaps, fearing for her safety, she didn't resist etc). Rather than view inconsistency as a natural or objective incongruity between contradictory facts of evidence we envision it as a contextually situated and multimodally emergent naturalizing process, emerging in part from the rhythmic integration of gesture and speech. If one definition of power is the ability to make one's account count (Giddens 1979, 1984) then interdig-

---

[41] Needless to say, this notion of multimodal ideology of language derives from and builds on Silverstein's classic work on language ideology (Silverstein 1976). Matoesian (2001:38) developed the notion of a linguistic ideology of inconsistency or that inconsistency is not merely a natural logic of contradictory pieces of evidence but a thoroughly infused logic of power and domination (at least on key occasions in the rape trial). Here we expand that idea to encompass other semiotic resources beyond speech alone.

ital beats in particular and multimodal conduct in general may offer a visual way to make it count.[42]

In the next chapter, we continue our exploration of closing argument, but this time with a focus on a novel type of interdiscursive resource designed to animate the texture of evidence in legal oratory.

---

[42] We are not making any claims about form and thus function of gesture as a one-to-one mapping of meaning. We do not claim that these function in a certain way in a specific context of use. Our aim is to analyze the gesture/speech produced in the here-and-now. We make no claims that a gesture/speech ensemble typically works in a certain way nor do we extend the analysis beyond their use here. As Adam Kendon and David McNeill have indicated there is no stable, conventional meaning of gesture and speech; their meaning is spontaneously produced in the moment.

6 *Unless he has three or four arms:* Enacting evidence

The past several decades have seen a surge of scholarly interest in reported speech in the law – and for good reason. The decontextualization and recontextualization (or interdiscursivity) of written and verbal language represent the evidential infrastructure of both the adversarial and inquisitorial systems. Researchers in forensic linguistics or language and law have studied such interdiscursive structures in numerous legal contexts (see Heffer, Rock, and Conley 2013). In her study of a cocaine possession case, Philips (1986: 154) found that direct quotes were a more truthful and reliable form of evidence than other types of reported speech: "quoting is reserved for information related to proof of the elements of a criminal charge, to foreground this information, and to give it more fixedness and credibility as 'exact words" than other forms of reported speech are given." Trinch (2003) studied how the protective order application interview of Latina victims of domestic abuse transforms women into institutionally relevant victims in the written affidavit. She noted how victim's stories of domestic violence are transformed and filtered through the institutional demands of the written legal report so the judge grants the restraining order. In her study of police interrogations, Komter (2019) examined how the suspect's statement circulates from the police interrogation to the written report and case file and finally to the trial. She demonstrated how interrogation talk is channeled into a legally relevant and authoritative document, a document produced with an evidential eye toward its future use in the trial. And in her classic study of law school socialization, Mertz (2007: 61) revealed how precedent, the cornerstone of the adversary system, links the historical to current case, building "analogies between the case before them and earlier cases."

Along a more practical vein, trial advocacy texts emphasize the role of interdiscursivity in trial practice. In his classic text, Mauet (1996: 247) recommends that when attorneys impeach witness credibility using prior inconsistent statements, they should "use the actual words of the impeaching statement." In a similar vein, Haydock and Sonsteng (1990: 45) note how "the goal of the trial attorney is to present the events in such a way that the fact finders think they are part of the case, perceive they are observing what actually happened in the past, and feel the emotions of the situation."

Legal actors mobilize speech and documents from historical context to create inconsistencies and impeach credibility in the current speech event. Reported speech like direct quotes, for example, constitutes an epistemologically and ontologically privileged form of evidence that captivates and engages an audience, creating a vivid image of legally relevant issues. Such "exact wording" makes testimony "come alive" and allows the jury to relive the actions, emotions, and events of the reported context. Thus, interdiscursive connections are crucial not only in an evidentiary but also emotional sense. Indeed, it is almost impossible to study the legal order without analytic attention to the nature and logic of interdiscursive structures and processes.

But there is another semiotic modality that figures prominently in interdiscursive connections, another semiotic resource that circulates between the historical and current speech event. Speakers do more than merely report the spoken or written word from historical context. They use their bodies to enact and demonstrate prior actions, actors and events in the current speech event. They use gestures and other embodied modalities to demonstrate prior moments of legal relevance and the evidential task at hand. While a number of studies have examined the legal relevance of reported speech and written documents few if any studies have shown how gestures and other embodied modalities function in the entextualization of evidential context. In this chapter, we examine how the attorney in closing argument does quite more than merely report historical statements; he enacts or demonstrates them in embodied form to provide a more captivating experience for the jury.

6.1 Demonstrations of evidence

According to Clark and Gerhig (1990: 765) demonstrations, like direct quotes, "work by enabling others to experience what it is like to perceive the things depicted." "They demonstrate what ... it looks, sounds, or feels like to perceive an event, state, process, or object ... to reenact or revivify it" (1990: 794). In contrast to descriptions that merely tell what happened, demonstrations involve a shift in perspective to show what actually happened. They give the audience epistemic access to the event depicted, an iconic re-presentation that draws recipients into a state of heightened involvement with and engagement in the events depicted. In terms of multimodal conduct, pure demonstrations not only depict historical speech a la direct quotes, but also and simultaneously re-enact the co-occurring gestures, gaze, and movements of previous characters and events, creating an authoritative and objective stance toward the events depicted.[43]

---

[43] For example, in their study of police focus group interactions Gilbert and Matoesian (2021: 61) described how one of the officers re-enacted the way small town citizens addressed the mayor in his rural town. He shifted gaze toward an unoccupied space, produced the direct quote *Hey Bob* with noticeably marked stress and eyebrow flash, and waved toward the imaginary mayor.

However, as Sauer (2003: 226) indicates, speakers can express two distinct viewpoints simultaneously, one in speech as a description, the other in gesture to reenact or demonstrate events as characters. In the segment analyzed here, the defense attorney produces a blended narrative of speech description and bodily quotation, the former telling the jury what happened in the previously occurring event, the latter showing them by animating bodies in the incident. Together, both speech description and embodied demonstration provide a richer representation of meaning than either could provide alone, allowing the jury to experience the event in the most engrossing way possible.

In what follows we analyze two segments of closing argument, the first describes the victim's statements to the police about male physiology during the rape, the second demonstrates the sexual anatomy – virtual positioning and movement of the bodies – of the incident through gesture while simultaneously describing it through speech. Together, the dynamic interplay between speech and bodily quotation elicits heightened displays of engagement and attention from the jury (Sidnell 2006: 405). As we will see below, the attorney organizes his closing argument multimodally around male physiology in the incident based on the victim's historical statements.

6.2 Anytime we talk about sex it's embarrassing

Example 1 Defense Attorney Roy Black's Closing Argument

```
01    Anytime we talk about sexual matters (.) it's embarrassing (.) and it's embarrassing
02    in this courtroom to talk about (0.3) But it's something that has to be discussed (.)
                           [((B 2H))   ((B 2H))      ((B 2H))    ((B 2H))
                           ((2 hand open palm supine beats=B 2H))
03    because of the allegations (.) that were made (1.4) Let me read to you what's on:::
      ((B 2H)) ((B 2H))                              [[((moves to fetch transcript))
04    (.) the questionnaire (4.2) This is a questionnaire filled out by::: miss Bowman
                            [((walks back to podium))
                            ((reading transcript as he walks back))
                            ((flipping pages))
05    followed up (4.5) with statements (0.5) on page five. This questionnaire is in
                       [((gaze to jury)) [((gaze to transcript))
                       ((reaches relevant page))
06    evidence as defendant's exhibit forty-three (0.6) You can- when you get back to the
      [((gaze to jury))              [((reaches podium; 11.1 seconds to fetch transcripts
                                       and return to podium))
07    jury room you can read it for yourself. You don't have to rely upon what I tell you
```

08 (1.0) She checked that the offender- (0.9) The question is was the offender able to
    [((reading from transcript in right hand)) ((LHB))   ((LHB))   ((LHB))   ((LHB))
                                        ((left hand chopping beats=LHB))
09 obtain and maintain an erection sufficient for sexual intercourse. The answer is
    [((LHB))   ((LHB))   ((LHB))   ((LHB))   [((transcript beat with right hand))
                                        ((returns gaze to jury))
10 **NO::::** (2.0)
    [(((transcript beat w/high elevation on upstroke+increased acceleration on downstroke))
11 Underneath that (.)
    [((traces hand across and gazes at transcript))
12 did the offender have a   *partial* erection?   (.)   **Yes**
    [((beat with left hand))   [((simultaneous beats with transcript and left hand))
    ((gaze at transcript))   ((gaze returns to jury))
    ((2 lines omitted))
13 (4.0) The testimony of doctor Good tells us (1.0) that if Patty Bowman was penetrated
              [((B))   ((B))   ((B))   ((B))   ((B))   ((B)) ((B))   ((BHE))
14 (.) with a *partially* (.) *erect* (.) *penis* (.) it had to be a *voluntary consensual act.* (1.2)
         [((B))   ((B))   ((B))   ((B))   ((B))   ((B))   ((B))
         ((right hand chopping beats = ((B)) ((high elevation beat = BHE))
         ((stress in line 14 indicates staccato))
    ((1 line omitted))
15 She had to *help.*
              [((2 hand open palm supine beat. Upstroke on *to*-particle. Downstroke on *help*))
16 She had to be *receptive.*
              [((2 hand open palm supine beat. Upstroke on *to*-particle; downstroke on
                receptive))
              ((short hold on *be* upstroke))
17 She had to *assist* the male in order for there to be penetration under these circumstances.
              [((2 hand open palm beat. Upstroke on *to*-particle; downstroke on *assist*))
((Lines 15–17 beat upstrokes to chest level, to waist level on downstroke))

In line 01, the defense attorney repeats the affective adjective, *embarrassing*, to modify a forthcoming discussion of *sexual matters*, a rather delicate topic not only *anytime*, but even in *this courtroom*. He then marks the contrastive parallelism grammatically with the coordinating *but* to indicate that, despite being a rather sensitive topic, it is something necessary to *discuss* because of the victim's *allegations*. Moreover, he layers contrastive parallelism with cross-modal repetition through a series of two handed, open-palm supine beats (palms up with lateral separation of the hands moving up and down). As we have seen previously, beats orchestrate the rhythms of speech and highlight significant information for the jury. When combined with open palms facing upward they intensify and foreground the proposition as obvious. That is, the need to discuss the delicate sexual matter conveys epistemic stance: a conviction that is unquestionably certain (Calbris 2011: 235–6; Cooperrider et al 2018: 8).

Yet, rather than merely report sexual matters, Black leaves the podium to fetch a police transcript containing an interview with the victim. It takes over eleven seconds to get the transcript and return to the podium. While walking back he flips through the pages, suspending utterance progressivity several times till he reaches the relevant page (notice the 4.2 second pause in line 04, and 4.5 second pause in line 05 as well as the vowel stretch on *by:::* in line 04 and *on:::* in line 3) and then returns gaze to the jury on line 05. Such a transparent display of involvement with the transcript reveals the dynamic interplay between speech and material conduct: how walking with and inspecting the transcript disrupts utterance progressivity until the hitch is resolved. More generally, it displays how material conduct figures in the pace and timing of his utterance in progress.

After returning to the podium something interesting happens. He first indicates the exhibit number (*forty-three*) and, second, directs the jury to inspect the transcript for themselves as *co-readers* of the source of information (*when you get back to the jury room* in lines 05–06) rather than merely take his word for it.

But why go to some lengths to fetch the transcript, riffle through the pages, and then direct the jury to read the transcript themselves, especially when he quite obviously knows the passage to convey? Why not merely report the relevant portion? Once again, according to Goffman's (1981) classic dissection of participant roles in conversation, the animator merely conveys the words, while the author writes those words, and the principal bears responsibility for those words. When Black reads from the transcript he positions himself as the animator while Bowman represents the author and principal, and in so doing he imparts an evidential aura of objectivity and authenticity to the transcript as if it were immune from bias in any way, elevating it to a state of epistemic privilege beyond reproach and distancing himself from any residue of authorship. At the same time, by using the official police document he sanitizes his forthcoming remarks as a necessary legal rather than sexual matter in a deft display of impression management: a display mindful of the tender sensibilities of the jurors.

Bowman's statement circulates from the interview transcript/police statements to courtroom testimony and finally to closing argument, a vivid illustration of how written documents provide the interdiscursive infrastructure of the legal order. While often considered separate lexical and grammatical concepts (Aikhenvald 2005), evidentiality and stance in the transcript simultaneously encode the source of information (evidentiality) and convey degree of speaker commitment to and certainty of the proposition (epistemic stance). That is, when considered multimodally via material conduct the source and value of knowledge co-occur simultaneously in the victim's own words. Her interview statement is taken as a factual, authoritative document, a linguistic ideology that first hand reports possess greater truth-value than other accounts, and then reproduced in court with an eye toward the evidential work it may accomplish, such as inconsistency (as we shall see). In fact, the attorney seems to say: 'you don't have to believe me that this is so unbelievable.'

The transcript contextualizes and frames Black's forthcoming reading of the *embarrassing sexual matter* from line 01, a matter that emerges on line 08. He begins with her answer but then cuts-off and recycles his utterance in progress to read the question verbatim, maintaining the integrity of the reported interview sequence and leaving its evidential value and morphophonemic repetition (reduplication) intact: *She checked that the offender- The question is was the offender able to obtain and maintain an erection sufficient for sexual intercourse.* The question is accompanied by a series of short left hand (vertical) chopping beats while reading the transcript in the right hand, once again emphasizing the information. On the turn-final NP (*for sexual intercourse*), he returns gaze to the jury and shifts to beat with the transcript with the right hand. The *NO:::* (line 10) co-occurs using a second transcript beat with the right hand but this time with high elevation on the upstroke and increased acceleration on the downstroke (see Figure 6.1 below). Increased loudness, stress, and vowel prolongation on the marked negative align perfectly with the marked beat to intensify and foreground a crucial piece of evidence about male physiology. Black beats out the significance of her words with her words. In line 11, he traces his finger across the transcript for the next question and answer in line 12 (*did the offender have a partial erection? Yes*) to further display that he is reading the victim's answer from the transcript itself rather than using his own words. And for additional emphasis, he beats simultaneously with both the left hand and transcript in right hand to intensify his stance that co-occurs with *partial erection*. Black claims that since the defendant had a partial erection the sexual

Figure 6.1: line 10 (*No::::*).

assault would be impossible and invokes expert testimony via the *if*-conditional to bolster that claim (in lines 13–14): *it had to be a voluntary consensual act.*

In lines 13–17, his narrative reaches a crescendo, both in form and meaning. If the defendant was only *partially erect* then the event had to be a *voluntary consensual act*. Just as important, the act was not only consensual but something that could only occur with the assistance of Bowman, a crucial piece of evidence that occurs in a densely laminated form of cross-modal text-metricalization. Such similarity driven and perceptually salient patterns of recurrence apply to co-speech gestures or multimodal structures as well as speech. As mentioned previously, for Silverstein (2014) text metricalizations like parallelism and other poetic forms function as metapragmatic guides to calibrate interpretation and evaluation of the narrative text. Most impressively, they deliver an aesthetically charged emotional effect, a multi-layered evocative rhythm that adds affective depth and intensity to Black's oratory.

With these points in mind, the poetic structure in lines 15–17 unfolds as follows. Each begins with the third person *she*, followed by the past auxiliary *had*, and ends with the stressed *to*-infinitive: *She had to help, She had to be receptive, She had to assist*. Each line co-occurs with rhythmic patterning not only in speech but in the integration of speech and gestural beats during the infinitives. The beat upstrokes co-occur with the *to*-particle, the down strokes on the stressed infinitives (*help, receptive,* and *assist*), with a short micro-hold on *be* in line 16 to maintain the rhythmic integrity of the poetic pattern (just before onset of down stroke). Finally, each beat occurs as a two handed supine gesture (open palm facing upwards or at an oblique angle) to intensify the evidential point. Reaching an evidential crescendo, Black foregrounds form over content in a dense metricalization of both speech and gesture – multimodal parallelism – as each reflexively elaborates the other in the rhythmic patterning of persuasive oratory. In so doing he drives home the evidential point about the victim's statement (see Figures 6.2 and 6.3 below). Her account is inconsistent with the facts – her own facts!

Still, while the official report of the victim's statement describes male physiology in the rape incident, it does not *show* us the details of that incident and the precise manner in which it occurred. In the next segment, the defense attorney *demonstrates* the manner in which the rape event occurred through a dynamic interplay of speech and gesture. That is to say, he re-enacts sexual anatomy – positioning and movement of the defendant and victim's bodies – in the rape incident based on Bowman's statements. He inhabits bodies in the incident through gestures while using speech to instruct the jury how to interpret and evaluate the demonstration. Black's blended viewpoint allows the jury to visualize dynamic processes that cannot be easily accommodated through speech, especially space, manner and motion, allowing them to experience the event in the most captivating fashion possible. As we will see next, there is much more interdiscursive evidence involved in the incident than just male physiology.

Figure 6.2: line 15 (upstroke on *to*).

Figure 6.3: line 15 (downstroke on *help*).

6.3 He does not have enough arms and hands to do that

Example 2 Defense Attorney Roy Black's Closing Argument

```
01    And think about for a moment (.) the description of what happened (3.8) Will
                                       [((BPF))      ((BPF))((BPF+hold))((B))
      ((BPF=2 hand beats with open palms facing one another))
02    is    supposedly on top of her (0.6)
      ((B)) ((B))          ((B))   B=right hand chop beats
```

03   His right hand (0.5)
     [(((start to extend right arm))
     [(((right hand partially closes in a grasping object gesture))
04   is   holding   her   left   arm
     [[((B))   ((B))   ((CG))   ((CG))
        ((B=partially closed grasp beats, loosely furled fingers of right hand))
        ((CG=closes grasp in a fist like gesture at full arm extension))
05   (1.2) Her other arm (1.2)
          [(((extends left arm to front of body w/open palm vertical
             while keeping right arm/hand held in extended grasping position))
06   is trapped between (.) the chest of the two of them
     [(((moves left hand to touch chest; on *between* 2 motion gestures to chest))
07   (0.5) His other hand is down at his side (.)
     [(((moves left arm straight down to side of body while still extending right arm/hand
        in grasping gesture))
08   He has to press his chest down to keep (.) her arm trapped between them (.)
     [(((retraction of right arm/hand grasp extension))
     [(((flat right hand/palm facing to chest and moves upper torso forward))
                         [(((three two handed beat gestures w/palms vertical
                             and facing one another moving outward from body))
09   If he raises up in any way (.)
     [(((left hand/arm flat on chest. Right hand grasp and arm extension repeat. Steps back))
10   that arm is going to escape and be able to attack him (2.9) And this whole time
     [(((left arm moves outward and to the side; retraction of right arm extension))
11   (.) she is struggling (0.7) She is trying to prevent in anyway possible any kind of
           [(((LCB))   ((LCB))   ((LCB))   ((LCB))
12   penetration (.)
     ((LCB)) ((LCB=lateral chopping beat with right hand))
13   At the same time he's holding down these two arms, one with
     [(((camera moves away from defense attorney))
14   his chest, and one with his hand (0.5) He has- His other arm is pulling up her
15   dress (0.5) He then has to hold her *panties* aside at the same time (0.3) **Hold**
16   **them aside** (1.6) and al:::so (1.0) somehow get his semi-erect or partially
17   erect penis into her vagina without her assistance (1.4) He simply does not have
18   enough arms and hands *to do that* ladies and gentlemen (1.7) That description
19   is ***impossible*** (2.1) Not only that (.) but for there to be actual penetration there
20   has to be the right angle involved (1.5)   None of that can occur with him
                           [(((camera returns   [(((right hand flat across chest))
                             to defense attorney))
21   pushing
     [(((moves upper torso forward w/head reaching end of podium))
22   down on her chest to keep her arm pinned between them
                           [(((punching gesture or right closed, fist to chest))
                             ((thumb facing chest on punch to chest))
                             ((dorsal side of hand up or horizontal))
23   (2.5) Unless he has three or four arms (0.7)
     [(((open palms supine with both arms spread+marked hold over entire line))

24	this act
	[((palms-open/oblique angle facing one another))
	((two-hand forward shoveling gesture to transcript))
	((gaze to transcript))
25	could not occur.
	[((gaze to jury/hands on podium))

After the assault, Bowman gave statements to the police and later to the prosecuting and defense attorneys. Moreover, during both direct and cross-examination, she gave the same description of the sexual anatomy of the rape incident that Black starts to address in line 01.[44] He begins with the dubitative stance adverbial, *supposedly*, that not only encodes a degree of skepticism about her description but also conveys a sarcastic attitude toward it (Nordlund and Pekkarinen 2014: 56). So from the start he entertains reservations about her description. In lines 03–06, he describes the defendant *holding* down Bowman's left arm while keeping her other arm *trapped* between both of their chests. His other hand is down by his side, While not formally acknowledged, Black authors the *if-then* conditional in lines 08–10: the defendant must *press his chest down to keep her arm trapped between them* because *if he raises up in any way that arm is going to escape and be able to attack him*. Thus in addition to the victim's description he inserts his own hypothetical into the case: what would happen if the defendant moves his chest up. In so doing, Black makes it sound as if Bowman could victimize the defendant!

However, while description tells jurors what happened it does not show *how* it happened in concrete detail. Speech does not easily accommodate the physical particulars of manner and motion involved in the sexual anatomy of the incident. Verbs like *trapped, press,* and *holds* are difficult to represent in speech. For example, is the defendant holding Bowman's left arm down to the left side of her body, downward vertically, or extended outward from the body horizontally? His right hand is "holding" down her left arm but how and where is he doing that? Phrasal verbs like "hold down" are spatially and directionally underspecified.

By reenacting the incident using gesture, Black reveals precisely where and how the defendant held down the victim's arm and trapped her other arm between their chests. First, he positions Smith and Bowman facing one another, with the latter on her back facing up, the former on top of her. Second, Black positions himself as inhabiting the defendant's body to demonstrate the spatial coordinates of the incident. When he does so, we see Smith's right hand holding down the vic-

---

[44] The two handed open palms do more than here than beat out rhythm (in line 01); they also possess a metaphoric function, presenting an abstract object for inspection (Mueller 2004; Calbris 2011).

tim's left arm. And, third, given that Black's arm is fully extended vertically we can assume that Smith holds her left arm somewhere over her head; that is, held down vertically extended upward above the head.

But not right away. Prior to the iconic grasping gesture, Black recruits two beats with the fingers loosely furled for emphasis, synchronized with *is* and *holding* in line 04. His beats unfold in precisely spaced increments until reaching full arm extension, where they evolve into the iconic grasping or closed hand gesture. More technically, they represent pre-strokes in route to the iconic grasp. That is, on *left arm* he closes the loosely furled digits into an iconic display of holding or grasping an object, in this case the victim's left arm (see Figure 6.4 below). Thus words describe the defendant holding down the victim's arm; the iconic gesture demonstrates where and how he is holding it down. Together, both speech description and gestural demonstration provide a more robust dimension of meaning than either could produce alone, enabling jurors to experience the incident as if they were present in the historical moment.

Figure 6.4: line 04 (*left arm*).

Black holds the iconic gesture in place, and after a 1.2 second pause extends his left arm vertically in front of the body to indicate the victim's right arm (line 05). Next, he moves his left hand palm flat on his chest to position where the victim's right hand is *trapped* between the two chests (lines 05–06, see Figure 6.5 below)). He then recycles the flat hand and moves it forward to his chest two times on *between*. The gesture and movement demonstrate metaphorically where and how her right hand was trapped. Put another way, Black recruits the flat left hand to inhabit a space for the chest and the two motion gestures for both chests (movement toward one another), using the gesture to inhabit two bodies simultaneously while still maintaining the grasping gesture with full arm extension till resolution of chest demonstration (in line 08). Unlike speech description, gestural re-enactment of trapping-in-motion demonstrates the *process* not just the outcome of the historical event. And in line 07, he states that the defendant's *other hand is down at his side*, and from the gesture we can tell it is his left hand, once again still extending the right arm and grasping gesture.

Figure 6.5: line 06 (*trapped between*).

Let us return to the insights of Jakobson and Silverstein discussed previously and examine in detail the polyrhythmic weave of poetic patterning in lines 03–07, a pattern that unfolds as follow. Black produces 1. a repetitive pattern in lines 03 and 07: *his right hand* and *his other hand;* 2. embedded repetition in lines 04 and 05: *her left arm* and *her other arm;* and 3. contrastive and displaced repetitions in lines 05 and 07: *her other arm* and *his other hand*, and lines 03 and 04: *his right hand* and *her left arm*. In the above lines, he produces intricate off-kilter repetitive rhythms that integrate re-enactment and description, a densely laminated form of text-metricalization, a finely executed expression of speech, gesture and imagery that packs a powerful emotive punch. More theoretically, Black's metrically driven and emotionally evocative pattern constitutes a design feature of affective oratory.

In lines 08–10, Black describes a hypothetical matter when Smith *traps* the victim's arm between the two chests. But once again, most of the action involves a gestural demonstration that co-occurs with the description. First, he retracts the arm extension grasp and redeploys his right hand flat over his chest (with palm facing chest) and moves his upper torso forward to iconically enact *press his chest down*, demonstrating how the defendant pins the victim's arm between the chests (see Figure 6.6 below). He then employs several two handed beat gestures to accompany the infinitive compliment. Second, in line 09, he places both his left arm (radial side of forearm bent at elbow) and flat palm of the hand to face his chest and reactivates the right arm extension grasping gesture. And third, as he does so, he takes a step back (see Figure 6.7 below) and moves his left arm outward and to the side, indicating how the victim's arm could escape if Smith raises up in any way (and withdraws the right arm extension during the 2.9 second pause in line 10). At a finer level of granularity, Black still has the right hand in a closed grasp and puts his left open hand (with the palm facing the chest) over his chest and moves back which is iconic of raising-up. On *if he raises-up* he moves backward to illustrate gesturally what would happen if her arm were not pinned. Once again, Black employs the conditional to demonstrate his own voice embodied, not a description from the victim, switching footing from describing and enacting the victim's testimony to describing and enacting his own hypothetical. His blended narrative of speech description and embodied enactment blurs the line between reporting and reported context by positioning his own voice alongside the victim's. And we will see how the conditional transacts important evidential work in a moment.

Figure 6.6: line 08 (*press his chest down*).

For now, a further point warrants consideration. The right arm extension with hand grasp shows how the defendant simultaneously holds down the victim's right arm while engaged in other actions relating to sexual anatomy. Black maintains the grasping gesture while describing how the victim's arm is trapped (lines 05–06) and how the defendant's other arm is at his side (line 07). He retracts the right arm extension in line 08, restores it in line 09, and then retracts it again in line 10. Such actions occur simultaneously with the upper torso and arm movements but without speech description. Black only conveys removing and retrieving arm exten-

Figure 6.7: line 09 (*if he raises up*).

sion through gestural demonstration. Indeed, these dynamic, oscillating actions are difficult to represent as economically and accurately through speech alone.[45] In a more methodological vein, without empirical consideration of gestures, such actions would go unnoticed and erased from analytic view.

Just as important, Black states that the victim is struggling and trying to prevent penetration, though without specifying her actions in this regard, and he uses a repetitive structure for emphasis: *she is struggling, she is trying*. Moreover, while the defendant holds down the victim's arms he has another task confronting him. Since the victim is fully clothed he has to *hold her panties aside at the same time. Hold them aside.* Black repeats phrasal verb (*hold aside*) this time with not only marked stress but increased loudness as well to augment the inconsistency in progress. The defendant does *not have enough arms and hands to do that*, making the victim's representation ***impossible***. (line 19).[46]

---

[45] It is not till lines 13–15 that Black officially announces that these gestures occur simultaneously.
[46] As the reader can see, on line 13 the camera moves away from the speaker, and that offers the opportunity to make a comment on the use of data in courtrooms. Audio-video recordings provide researchers with unprecedented access to visual data. However, audio-video data, like all data, is far from perfect, and while indeed valuable for enhancing the precision of empirical research, there are practical and logistical considerations that figure in the use of multimodal data in legal settings like courtrooms. Frequently, video recordings of the proceedings are not permitted (at least not in the United States nor in the United Kingdom) and even in those cases where they are, it is quite different from videotaping everyday conversation where researchers can position the camera angle or even deploy multiple cameras to include the entire set of participants. Instead, researchers have to rely on third parties (like various news sources), which means that the available visual record will be selective in what is recorded and, occasionally, all engaged participants and

Thus, the defense attorney augments the issue of male physiology in several steps: first, the victim is struggling; second, the defendant uses one arm to hold down the victim's left hand; third, he pins her other hand between their chests; and finally, he has to hold her panties aside to complete the sexual assault. And he is not finished.

Let us return to the conditional in line 09–10 and pose the question: Why does Black shift footing from a demonstration and description of the victim's report to his own hypothetical in a constantly modulating participation structure? Why not continue to animate motion of the bodies in the victim's voice? He uses the shift as the evidential platform to launch another demonstration – a final increment to the sexual anatomy of the incident and the striking inconsistency in it: the necessary *angle* for the sexual assault to occur (in lines 19–22).

In this demonstration (and for a second time), he places his right hand palm flat over the chest and bends his upper torso down and forward towards the podium in noticeably marked fashion (and looks like he does a "pushing" motion at the same time, see Figure 6.8 below). More explicitly, on *right angle* in line 20 Black positions his right hand as Bowman's left hand that is trapped between the two chests as he moves his upper torso forward. He then recalibrates his right hand into a fist gesture with the dorsal side up (lateral knuckles facing up or horizontal knuckles of closed hand facing up laterally) in a hitting-the-chest or *pushing* motion to demonstrate how he *keeps her arm pinned* (see Figure 6.9 below).[47] Thus the defendant and victim's bodies are not properly aligned for penetration to occur.

We can now see why the grasp is so important to keep in visual-evidential play, why Black contingently retools it to deal with other relevant particulars of motion. In line 04, he mobilizes the present progressive *is holding* to show that the action is ongoing or continuous. However, while speech fades rapidly (one of Hockett's classic design features of course), manual gesture is able to show how the extended right hand is continuously occupied, allowing the jury to not only see the lack of alignment but several co-occurring actions as well (something speech cannot do with the same level of economy and efficiency). Put another way, gestural aspect (as it were) remains visible and in play well after its fleeting speech counterpart has disappeared. Progressive aspect describes the continuity while the manual gesture allows the jury to visualize that continuity in motion.

---

relevant interactive contours of multimodal conduct may not be captured on the video or, as in the above case, the camera angle moves away at random.

47 Black positions the thumb (in line 22) as pressed between the chest and rest of the fingers of the closed hand (in the hitting motion) as metaphoric of another chest or the thumb pinned against the chest (that is, his thumb is pressed again his chest rather than thumb up). If the fist were vertical or thumb up rather than horizontal this metaphoric imagery would not emerge.

Figure 6.8: lines 21–22 (*pushing down on her chest*).

Figure 6.9: line 22 (*to keep*).

Black's narrative reaches a crescendo in lines 23–25 with the *unless* conditional, and here he mobilizes first a lateral open palms facing up gesture (line 23, see Figure 6.10 below), once again imputing an obvious point, and second an open-palms gesture with the two hands at an oblique or slanted angle facing one another (on the demonstrative *this act*) to produce a *shoveling* dismissive gesture (line 24) toward the transcript on the podium (see Figure 6.11 below). At the same time, he gazes at the transcript as if pointing to it with his eyes, thus drawing the jury's attention to and establishing the transcript as a joint focus of attention. In so doing, his gaze turns the gesture, once again, into an interactionally significant object, a transparent display of engagement with the transcript and the inconsistencies embedded in it. He concludes by placing both hands on the podium and returning gaze to the jury box.

Figure 6.10: line 23 (*unless he has three or four arms*).

Figure 6.11: line 24 (*this act*).

In this section, we have seen how one can describe, rather statically, use of the arms and hands or, much more dramatically, re-enact the positioning and movement of hands, arm, and body as evidence that resonates with the jury. Gestural re-enactment possesses a subliminal facticity that naturalizes the characters and events depicted in it. Black's interdiscursive strategy foregrounds each gestural segment and galvanizes the authority of gesture to shape our perception of legal-evidential realities. Embodied re-enactments dissect the incident into discretely layered segments of manner, motion, and direction unfolding in a linear sequence, a sequence that allows the jury to re-experience the action in a way that mere words alone may not accomplish. For example, the oscillating hand grasp allows him to keep one segment in play (or remove it) while adding additional ones. In sum, Black uses speech to describe the inconsistency; gesture to demonstrate and magnify it, enabling the jury to *re-experience* the evidence *in motion*. Together, speech description and bodily quotation provide a richer representation of meaning than either could provide alone, stimulating the jury's attention in the process. As an anecdotal proof (but proof nonetheless), Black stated in a taped interview Matoesian conducted with him after the trial: *one of the jurors said after the trial that she tried that with her husband and without assistance it was impossible.* [Tape 1, side B about half way through tape.]

6.4 Summary

In conclusion, we wish to shift direction and entertain a much different perspective on Bowman's inconsistency. As we have seen, Black's closing focuses on inconsistencies between the facts she conveyed on the one hand and defense expectations about sexual performance given those facts on the other.[48] But what is inconsistency? How does it work? And why is it so powerful in constructing credibility in cases like the current one?

Inconsistency is inherently interdiscursive, a linguistic ideology – hegemonic rationalizations about language in particular and multimodal conduct more generally – that decontextualizes evidence from historical context and recontextualizes it in a current speech event.[49] Rather than view inconsistency as a logical incongruity among aspects of evidence or a generic system of logic, which is how it is viewed in the law, we consider it a logic of power embodied in the interdiscursively segmented interplay of multimodal signal streams in the construction of legal context. We conceptualize inconsistency as a linguistic ideology for generating inconsistencies rather than just uncovering pre-existing ones through erasure of sociocultural difference, at least in the case under consideration here. As Gal and Irvine 2019: 20) put it: "*Erasure* is that aspect of ideological work through which

---

[48] In more technical-legal terms, the inconsistency is between the victim's statements and the expert testimony of Dr. Good, who claimed that given Bowman's statements the act had to be not only consensual but also cooperative – with her assistance.

[49] As Woolard (2021:1) puts it, linguistic ideologies "are in some crucial way *about* language itself, rather than all ideologies encoded *in* and *through* language." They represent "morally and politically loaded representations of the nature, structure, and use of languages in the social world." We add that linguistic ideologies refer to not only language and speech but the ensemble of multimodal signals more generally.

some phenomena ... are rendered invisible ... goes unnoticed." Seen in this light, inconsistency represents a linguistic ideology that invokes and naturalizes hegemonic structures, concealing some sociocultural fields while noticing others. In this case, it foregrounds hegemonic male structure in and as adversarial cultural representations while ignoring the female experience of sexual assault and the ensuing trauma that emerges from it – gendered forms of experience rather than objectively neutral forms of legal reasoning.

More explicitly, inconsistency as a legal-linguistic ideology erases the post-trauma context in which Bowman's statements were produced. According to Burgess and Holmstrum (1974; see also RAINN 2008: 1) rape trauma syndrome in the acute phase (precisely when Bowman gave her statements) is characterized by shock and disbelief, poor recall of the assault, and disorganized thought more generally (or dulled memory function).[50] And that erasure of trauma becomes all the more glaring since the prosecution failed to have an expert testify on the features of rape trauma syndrome (and omitted from her closing argument). Thus what makes Black's oratorical skills so persuasive is not just the cross-modal demonstration but the hegemonic interplay between the linguistic ideology of inconsistency and his gestural re-enactment.

More theoretically, at key moments the linguistic ideology of *gendered inconsistency* appropriates the adversary system as the adversary system arrogates gender identity in a way that something novel emerges, an elective affinity between both institutions that conceals gender identity and naturalizes it as the adversary system with its system of generic inconsistency and gender neutrality in the process. Put another way, the adversary system erases inconsistency as a gendered object, concealing it under its auspices. As Woolard, (2021:3) mentions this is how social institutions become endowed with linguistic authority as they "regiment inequality, such as ... courts of law."

Thus, Black's multimodal enactment shapes our perception of inconsistency in the incident, positioning it beneath the limits of awareness (Silverstein 1981). Gestural reenactment works in concert with speech to naturalize inconsistencies and conceal how the "naturalization process" functions in microcosmic detail – how it is constructed.[51] Thus rather than conceive of inconsistency as some natural juxta-

---

50 In her study of rape myths (from rape trials in England and Wales), Smith (2018: 66) writes in a quite similar vein: "Rape trauma syndrome of other elements of the psychology of trauma demonstrate that the behavior being portrayed as unusual is actually commonplace in sexual violence ... barristers created overly simplified depictions of rape and argued that the jury must look for evidence of inconsistencies between witness testimony and their expectations of what a 'normal' person would do."
51 All reported speech, including direct quotation, is constructed speech rather than an exact replica of historical context. If gestural re-enactments *demonstrate* or *re-enact* like direct quotes (Clark

position of opposite or contradictory facts, we suggest that its sense of naturalness and taken-for-granted facticity emerge in the dynamic interplay between gestural re-enactment and linguistic ideology.

---

and Gehrig 1990) then why assume they produce an accurate or faithful replica of the reported context any more than their direct quote counterparts? Like direct reported speech, bodily quotes function to not only carry denotative-referential meaning but also accomplish interactional work in the here-and-now construction of evidential realities.

7 Conclusion

The conclusion consists of two parts. The first part gives a summary of the chapters and major findings, the second shows the relevance of our findings (and previous findings) for recent recommendations for reforming the rape trial.

Oratory – the art of delivering a speech to persuade an audience – first appeared in the law courts of Athens centuries ago. But it was Quintilian in Roman court who provided the first systematic study of legal oratory during classical antiquity, even opening a school of rhetoric. In his classic *The Orator's Education* (volume 11.3), he inaugurated a quest for the "perfect" (or "ideal") judicial orator.[52] As he emphasized, the integration of improvisational gestures and other embodied resources with speech represents the necessary features of an emotional "delivery" to persuade or "move" (sometimes he uses the term "stir") the judge in legal cases (Quintilian 11.3 page 175). And as we have seen in the current study, defense attorney Roy Black might well be the *perfect* candidate to fulfill that search – at least in the modern legal order.

Given Quintilian's historical treatment of the role of gesture in legal oratory, we find it unfortunate that contemporary studies in forensic linguistics or language and law have, for the most part, ignored the role of gesture and other embodied resources in legal discourse. In so doing, a significant dimension of the law has been erased from empirical observation and analytic consideration. By the same token, the growing field of gesture studies has ignored the role of gesture in particular and multimodal conduct more generally in legal context. In this volume we have made a contribution to both fields by showing how multimodal conduct in the law represents a necessary feature for any comprehensive investigation of forensic communication and that the study of multimodal conduct in legal oratory can make a significant contribution in its own right to the field of gesture studies.

Since the outset, we have emphasized how the rape trial – like all adversarial trials – is not about truth and falsity but winning and losing, and that depends on who can best persuade the jury through not just speech but through the integration of speech, gesture, and other modal resources: through multimodal conduct. In the spirit of Quintilian's quest and Bulwer's dialects of the fingers and more recently the insights of Jakobson and Silverstein, we have analyzed cross-modal poetic structures – the rhythmic integration of speech and embodied modalities in parallel and repetitive structures – and their role in evaluating evidence and persuading the audience: a multimodal oratory designed to get the jury to accept one version

---

[52] Of course, Quintilian, was not only an educator of rhetoric but a practicing lawyer (or legal pleader as it was called) in Roman Court.

of reality rather than another. And we have spent considerable time examining the role of beat gestures in that capacity and have demonstrated that they are just as intricate and sophisticated as other – especially imagistic – members of the gestural quartet.

More specifically, we examined how these poetic patterns function in several under researched areas in the field of language and law such as objections, event structure, collective memory, and bodily quotes or enactments.[53]

Chapter 2 looked at the institutional reflexivity of questions, and how they function both retrospectively and prospectively. More explicitly, we saw how participation roles a la Goffman take contingent precedent over adjacency pairs and the "question-centric" model of institutional discourse. As he mentions, the question-answer pair may be inadequate in some contexts and here we have seen why. One cannot ask any question in court. The felicity conditions governing participation roles, rules of evidence, and legal procedure take precedent over the prevailing view that communication in trial exam consists primarily of question-answer pairs. Moreover, although the defendant is officially excluded from participation in the objection event, he uses embodied conduct as an "unofficial" evaluation of the prosecuting attorney's legal identity and evidential credibility. More technically he employs embodied stance as a residual form of participation in objection mediated question-answer sequences. In a similar vein, the judge mobilized gaze shifting movements to signal egregious evidentiary violations to the opposing attorney (as if prompting him to object), even though the judge is suppose to be a "neutral umpire" in the adversary system. Without an examination of multimodal conduct such socio-legal action would be erased and excluded from analytic consideration.

In addition to the problems mentioned in chapter 2 concerning the applicability of the question-answer pair framework in trial context, let us address several possible attempts to salvage the idea since it represents such a deeply entrenched assumption about legal discourse. One point that warrants consideration is this. While the question is directed only to the witness, it gets a response not a reply from another participant – the attorney. Second, the objection cannot be heard as a second pair part to a question. And third it might be argued that questions in court project two types of conditionally relevant pair parts, an answer by the recipient

---

[53] By highlighting the poetic function we do not wish to posit false dichotomies, that, for example, the poetic function outweighs content or vice versa. On the contrary, both form and content mutually elaborate one another to create an emergent field of persuasive evidence. For instance, as we demonstrated in chapter 5 inter and intra digital beats encode both rhythm and residual semanticity simultaneously.

and/or an objection by the opposing attorney. The problem with this idea is that most improper questions do not inherit an objection. As Durkheimian social facts, adjacency pairs involve a state of conditional relevance or given a first pair part a second is normatively required and if missing it is officially missing. Objections do not possess the same nature and logic. They are optional contingencies – a contingent enactment of an evidential felicity condition – that may or may not materialize, even in the case of an improper question.

Chapter 3 analyzed how the multimodal encoding of motion events on the surface of a material object transformed opening statement into an (accusatory) evaluative argument about witness credibility and inconsistency, even though opening is only suppose to "refer to anticipated evidence" (Mauet and Easton 2021: 74). We examined how the defense attorney mobilized iconic gestures to trace movements on an exhibit, including an acoustic gesture in which the jury could hear the tracing, and how an ensemble of repetitive gestures shaped a static exist verb into an agentive motion verb.

In chapter 4, we examined how opening statement turned into an emotionally riveting moment of collective memory. In this case, the defense attorney employed cross-modal poetics as a microcosmic ritual to expand the pool of co-victims that included not only the defendant but even members of the jury. We also saw during both direct and cross-examination how the witness and attorneys worked in concert via multimodal conduct and paralinguistic cues to link collective memory to the present trial context, recalibrating socio-legal identity into a sacred ritual performance.

Chapters 5 and 6 turned from opening statement to closing argument. Chapter 5 investigated how inter and intra digital beats functioned as cross-modal poetic patterns to shape the perception of jurors about inconsistencies regarding the victim's clothing. We developed Bulwer's dialect of the fingers to show how the attorney counts off and counts up inconsistencies in the prosecution's case.[54] Contrary to orthodox wisdom on beats, such gestures not only beat out rhythm but convey semantic content as well.

Finally, chapter 6 analyzed how the attorney used multimodal conduct to combine speech representation with an embodied enactment of the victim's testimony, providing a more vivid and captivating interdiscursive experience for the jury. Most impressively, by enacting or demonstrating testimony, the attorney keeps one gesture in play continuously, while rotating other gestures into position concurrently, oscillating movements difficult to capture in speech alone as efficiently and effectively. That is, he calibrates and recalibrates positioning of bodies for

---

54 In a similar vein, Quintilian (book 11.3: 143) also discusses "telling off arguments on our fingers."

interpretive effect, creating novel positional configurations to manage the evidential task at hand.

To repeat our main goals. First, we began our studies in an effort to make a significant contribution to forensic linguistics by demonstrating in concrete detail the value – indeed, the necessity – of including multimodal conduct in any analysis of legal discourse. And, second, we revealed how the study of multimodal conduct in the law can make distinctive contributions to the field of gesture studies. Building on the original insights of Jakobson and Silverstein, we demonstrated how the poetic function of language not only inheres in speech but in the cross-modal integration of speech and gesture; specifically we saw how beats not only possess a rhythmic foregrounding affect but also, when combined with syntactic parallelism in speech, produce a rhythmically balanced and emotionally evocative choreography of speech and gesture that may function, at certain moments, semantically, conveying substantive information in its own right. That is to say, the complex organization of inter and intra digital beats convey a degree of residual semanticity in addition to their more orthodox rhythmic foregrounding function. We also saw how tracing gestures consisted of "flinging beats" toward the material object that transformed event structure from a static exist verb to an agentive motion verb. One final – rather novel – gestural resource warrants repeating: we saw how the attorney crafted an acoustic gesture while tracing on the exhibit and the metaphoric-evidential meaning of such a gesture in the construction of legal context.

More generally, we have witnessed how rhetorical devices function in legal oratory as a persuasive performance designed to move the jury, to dramatize sociolegal meaning and to guide interpretation and evaluation of the recurrent elements. Quintilian's (book 11.3: 89) words are worth mentioning once again: "since words are very powerful by themselves, and the voice adds it's own contribution to content, and gestures and movements have a meaning, then, when they all come together, the result must be perfection."

7.1 Rape reform

Over the past twenty years or so a number of scholars have proposed significant ways to reform the rape trial process not only in the U.S. but also in England, Wales, New Zealand and other countries that employ the adversary system of justice. In this second and final section of the conclusion, we cover the most prominent (and interrelated) proposals for reform: (1) pre-recorded questions and a neutral translator for cross-examination, opening statement and closing argument; (2) improving the truth and accuracy of evidence; and (3) rape myths.

7.1.1 Pre-recorded questions and neutral translator

In her study of rape trials in England and Wales, Olivia Smith (2018: 166) advocates the use of pre-recorded questions during cross-examination of the victim, primarily because of the manipulative and aggressive cross-examination by defense attorneys. Attorneys would "submit their cross-examination questions in advance" for inspection and translation of "misleading stereotypes," "overly manipulative or intrusive questions," and leading (instead of open-ended) questions (Smith 2018: 183). Similarly, Anne Cossins (2020: 587; 599–615) in her study of sexual assault trials in England, Wales, and Australia proposes a "pre-trial hearing" where the defense would submit their questions in advance to a "specialist examiner" or intermediary who would determine their appropriateness for a "trauma-informed" system of justice. Opening statement and closing argument would also be "vetted" to prohibit inappropriate attacks on the complainant's behavior before, during and after the assault. The pre-recording would elicit the best evidence and vetting would eliminate "complex leading questions" and maximize the use of open-ended questions. While Smith and Cossins propose pre-recorded questions, Taslitz (199) advocates a neutral intermediary to translate the defense attorney's questions during cross-examination into "less abusive forms" and to remove obscure, vague, and ambiguous language, thus reducing the victim's trauma of being raped a second time on the witness stand.

However, several relevant issues emerge when considering such reforms – and not merely logistical ones.

(1) Good attorneys do not write their questions (or opening and closing narratives) out in advance (Roy Black personal communication; see Matoesian 2001). They build their next question off of the witness's answer, a much more improvisational than static conception of how attorney's actually operate in the trial.

(2) As we have seen in some detail, questions and narratives consist of much more than speech. Co-speech gestures and other forms of embodied conduct impart considerable information in the attorney's speech, information often unavailable from the verbal modality alone. In Chapter 2, the witness employed embodied stance to critically evaluate the prosecuting attorney's questions, and in Chapter 6 the attorney enacted characters to increase the vividness of legal performance and thus persuade the jury. How would such embodied conduct be "prepared in advance" or "translated?"

(3) Chapter Two revealed that the most bullying, aggressive, and improper questioning – not to mention unethical – in the entire trial involved the prosecuting attorney's cross-examination of the defendant. The prosecuting attorney not the

defense attorney, engaged in egregious and unethical violations of trial procedure and evidence when attacking the defendant.

(4) How would disputes about the translation or vetting of questions/narratives be handled and who would make decisions about the specialist examiner's translations? This is no small matter, for the scholars making recommendations for change use vernacular and value-laden glosses when describing the defense attorney's use of language, such as "aggressive," "bullying," "whacking the complainant," "manipulative," "intrusive," "repetitive," "discourteous," "humiliating," and so on (see Craig, 2021 for numerous examples). It is not transparent what such terms look like in the concrete details of real-time courtroom discourse, and who will decide if such classifications apply in a particular instance.

(5) This relates to another problematic issue. Because of courtroom constraints on the questioning of witnesses, accusations typically consist of an incremental and progressive build-up of facts, leading up to the main point or accusation (as Atkinson and Drew 1979 noted in their classic study). Moreover, questioning in court, like everyday contexts, is inferential, as we have seen throughout this volume. Where does the translation or vetting begin and end? Such logistical items are not tangential to reform but inhere in the nature of communicative practices, and reformers rarely if ever provide such concrete details for their recommendations.

(6) And related to the above points, without knowing what is "aggressive" or "manipulative" or "repetitive" and so on (vague notions at best), it is not clear what would be specific candidates for translation or vetting and when and where this would occur in the sequence. If reformers wish to make claims about language use and multimodal conduct in the rape trial (or any trial for that matter), which is what questions (or opening statements or closing arguments) actually are, then the unit of analysis must be multimodal conduct: words, utterances, gestures etc. To merely assert that this or that utterance is aggressive or manipulative or intrusive or irrelevant or repetitive is to employ vernacular glosses that can be "read off" of *a priori* advocacy assumptions about what happens in the rape trial. That is to say, reformers like Smith make the rape trial, not discursive practice, the unit of analysis. For example, a key feature – indeed the hallmark – of the poetic function in discursive practice is multimodal repetition, as we have seen throughout this volume.

Thus by making the rape trial as their unit of analysis, reformers base their recommendations on explicit and quite selective cross-examination segments of the victim only.[55] That is, they make selective use of data to prove selective advocacy

---

[55] And these explicit segments (such as those found in Craig, 2018) do little more than appeal to the prurient interest (see Matoesian and Taylor 1983 for a critique of this type of research).

assumptions, *a priori* assumptions about the rape trial and then select excerpts to "prove" those assumptions. And that gives us a narrow and quite misleading view of what happens in the trial. Put another way, we never see any analysis of the data but are merely shown data selectively to prove *a priori* assumptions about the rape trial in general and cross-examination of the victim in particular, as if the data merely "speaks" for itself. In this volume, by contrast, we have examined features of the rape trial in microscopic, multimodal detail. Rather than using multimodal conduct as an unexplicated resource we turned it into a topic in its own right. For example, Chapter Two revealed how courtroom examination is not a question-answer pair, but an objection mediated question-answer event. In our previous companion volume to the current (Matoesian and Gilbert 2018) and Matoesian (2001), we challenged the "question-centric" model of socio-legal discourse in court and demonstrated how answerers can recalibrate questions put forward by the attorney and how witnesses – the rape victim – may employ factive verbs that presuppose the truth of the embedded clause even under negation (i.e. after a defense attorney's question the victim responds with "you mean when he raped me?").

(7) Such recommendations for reform trade on the assumption that the credibility of the victim represents the key variable in the rape trial, and proposals for reform should reduce the harrowing degradation ceremony the victim endures during cross-examination. In fact, the authors mentioned above focus exclusively on the credibility of the victim as the most crucial variable for redressing the injustices of the trial. However, as we saw in Chapter Two, the credibility of the attorney not the victim or other witnesses may be the most crucial variable when prosecuting rape cases in particular and adversary trials more generally. As we saw, the credibility of the prosecuting attorney came under scrutiny and evaluation, perhaps more than anyone else in the case.

(8) Smith (2018: 183) and Cossins (2020: 607) argue against the use of leading questions during cross-examination of the victim, and instead advocate open-ended questions that do not limit the victim's voice. However, as Matoesian and Gilbert (2018) demonstrated in multimodal detail, defense attorneys find that in such cases the witness will quite frequently put their "foot in their mouth" as it were, and impart damaging information to the prosecution case. As Roy Black mentioned in an interview with Matoesian (see Matoesian 2001: 21; Matoesian and Gilbert 2018: 204), the victim's taped statements to the police "sounded like a session between a psychiatrist and patient more than a police officer and a witness to a crime." And those recordings were played in the trial, much to the dismay of the prosecution.[56]

---

[56] See Matoesian and Gilbert (2018) Chapter Four for another poignant illustration of this issue.

(9) And last, Taslitz recommends the use of a neutral translator because defense cross-examination uses obscure, vague, ambiguous, and convoluted language to "trick" the victim. However, Matoesian (2001) found some years ago that cross-examination, on occasion, was just the opposite. It was very concise, direct, and factual. Indeed, the trial practice known as detailing-to-death trades on those very premises. In any event, the translator will not be a linguistic magician, no matter how well trained, so the translation will never be objective but rather another branch in the tree of moral-inferential work, as is inevitably the case.

7.1.2 Seeking truth and the accuracy of evidence

Olivia Smith (2018: 197) advocates that reforms should "prioritize truth seeking" in the trial. Sue Lees (2002), Elaine Craig (2021), and Anne Cossins (2020) argue that the reforms should "purify" language to obtain the truth. They base their proposals on an ideology of referential precision and objective authority, a rationally instrumental view of language. Truth is something "out there" to be discovered if we can only get beyond the manipulative and distorting power of language.

As we have mentioned throughout this volume, however, the rape trial, like all trials, is not about truth and falsity but winning and losing and that revolves around who can best persuade the jury in and through multimodal oratory. As Michael Silverstein (2014) noted, this "dominance of denotation" ideology (rational propositions that consist of truth and accuracy) ignores how indexical and metaphoric meanings are primary not just "added on" characteristics of language. Instead, meaning is dependent on features of context in which the expression is uttered, and language cannot be purged of such stylistic features and purified to reach an objective and referentially precise realm of truth and accuracy.

Such a preoccupation with truth and accuracy of some state-of-affairs in the real world seeks to purify language and purge it of all "irrational," stylistic and aesthetic features of discourse – gross attempts to equate language with the referential function only. Yet, the referential or denotational function – language corresponding to some state of affairs in the world that can be determined to be true or false – is only one of the functions of language (Silverstein 2014: 129–130). As Bauman and Briggs (2003: 44) note, this eternal quest for a rational and scientific language is really an attack on rhetoric or the aesthetic use of language in particular and multimodal conduct in general (an attack on the poetic function or "verbal ornamentation" Bauman and Briggs 2003: 45): the indexical functions of multimodal conduct and poetic eloquence in discursive practice that we have investigated in this volume.

For instance, oratorical style and aesthetic features of discourse, like cross-modal poetics, contributes as much if not more to meaning in court as the content of questions, especially when it comes to attributing blame and allocating responsibility (that is, evaluation). One final example from this volume illustrates the point further. The co-production of a solemn paralinguistic sacred performance between the defense attorney and Senator Kennedy – lowering the volume of their speech in concert with one another – is not some tangential ornamentation to the co-construction of meaning but IS the meaning (from Chapter Four).

We have seen in this volume how repetition (and recall that one of the criticisms of reforms involves the overly repetitive nature of questioning) is not only the hallmark of persuasive oratory but also a response to constraints on questioning and evidence in court. Since everyday responses like "Give me a break" would be considered argumentative in court (and not a question) and thus objectionable, attorneys use repetition and other aesthetic devices to evaluate the witness's response. Text-metrical forms signal specific strategies that tell listeners how to interpret and evaluate the recurrent elements. How will the translation or vetting of multimodal conduct/language deal with such issues?

There may not be a single objective reality in the legal order to which the the truth can be ascertained from some Archimedean vantage point and thus distinguish truth and falsity. Reformers wish to change the law but without an appreciation that, in the words of Peter Tiersma (2000) in his classic *Legal Language,* our law is a law of words, and as we have shown here a law of multimodal conduct. If the law consists of the integration of language and gesture then any attempt to change the law must ultimately take multimodal conduct as a topic of study and not a mere unexplicated resource. Some rational form of logic and denotational truth simply does not exist – at least not in the social construction of rape's legal facticity. (this is not to say that truth does not exist but it may not exist in language). In a much more general theoretical sense, this concern with truth, accuracy and precision reflects the centuries old quest to separate *fact from moral value,* an impossible program (and the various proxies for this such as the legal/extra legal dichotomy).

One final item: Olivia Smith 2020, 2018 advocates the use of "handwritten field notes" and "prioritizes direct quotes" as data (Smith 2018: 8–9). However, if we return to Chapter 3 and the defense attorney's quote of the victim, "I've been raped come and pick me up", merely quoting content will not tell us what such utterances are doing in the construction and co-construction of legal context. As Silverstein mentions such attention to the narrow content or topic of talk – denotation-referential content – limits our ability to adequately interpret what that utterance is doing in context, what he refers to as the limits of awareness. As Gilbert and Matoesian 2021: 150 put it: merely quoting content or topic "foregrounds aspects of speech that are referential (describe some state of affairs), segmentable (morphosyntactic), and context presupposing in contrast to those that are more stylistic, nonsegmentable and context creating." They go on to mention that "By focusing on the topic or referential aspects of meaning" researchers omit those subtle contextualization cues like poetics, gesture, and the interactive aspects of meaning." In fact, just about the entirety of our volume focuses on the multimodal co-construction of meaning – most of which would be erased by "observing" or "quoting" the proceedings. Just as crucial, taking notes is not a substitute for audio-video recordings and detailed transcriptions of those recordings.

As Duranti (1997: 337) mentions: "Linguistic communication is part of the reality it is suppose to represent, interpret, and evoke." Put more prosaically, this eternal quest for the Holy Grail of communicative accuracy, referential precision, and truth will never, given the nature of the linguistic beast, likely succeed.

7.1.3 Rape myths

The final reform proposal involves the defense attorney's use of rape myths: "prescriptive or descriptive beliefs about rape that serve to deny, downplay or justify sexual violence" (Smith 2018: 55), such as "lack of injury/torn clothes," "failure to resist," "sexual history," "kissing as consent," "post-rape behavior," and so on (Tempkin, Gray, and Barrett 2018: 210). Cossins (2020: 184) defines them as "prejudicial stereotypes or false beliefs about rape, rape victims and rapists" and argues that the court should "prohibit questions based on rape myths" (Cossins 2020: 587–8).

Of course, the major myth is sexual history, and as such represents the primary candidate for vetting and translation, not only during defense cross-examination of the victim but also in "the content of opening and closing" (Cossins 2020: 597).

However, as we have shown throughout this volume, discursive practice in court is inferential and indirect, and if this is so practically anything can be construed as a rape myth based on sexual history, inconsistency, and any of the other myths. In fact, virtually anything considered unflattering to the victim or damaging to the prosecution's case can be construed as a rape myth. But as Matoesian (1995) demonstrated some years ago, the defense does not need to bring up sexual history directly to bring up sexual history. Sexual history simply goes through the inferential "back door route" of moral character, and such descriptions would likely surface even with further reforms.[57]

This is not to say that rape myths do not exist. Indeed, we have shown how they function in the microcosmic details of multimodal conduct, through linguistic ide-

---

[57] Even ordinary address terms like Miss Smith (see Matoesian 1993) and triangulation of kinship terms such as "your daughter's father" may bring up sexual history (see Matoesian 2001).

ologies in particular. But for the concept to have any empirical validity and analytic usefulness reformers must be explicit and specific about how they function in the construction and co-construction of legal context.

In sum, by looking at the microcosmic details of multimodal conduct in the trial we have demonstrated how our work makes a substantial contribution to the field of language and law or forensic linguistics on the one hand and gesture studies on the other. Just as important we have shown how the study of multimodal conduct bears considerable relevance for recent policy initiatives.

While it is customary to offer constructive proposals for reform at the end of a study like this, especially since we have critiqued other attempts, unfortunately we have little to offer – no workable solutions for reform policy or profound ideas to impart – other than this. Any attempt to reform the adversary rape trial or any type of trial for that matter must confront the rather transparent fact that the law consists of language and multimodal conduct in the construction and co-construction of legal context.[58] Ignoring how we *practice linguistics without a license* leaves us with an impoverished understanding of how the law functions.

---

[58] This is not to say that reform efforts to achieve justice are impossible but rather to stress that attempts must start from the multimodal infrastructure in and through which such reforms are applied. Matoesian (1995) demonstrated precisely this point in explicit detail some years ago.

Transcription conventions used

Symbol	Meaning
(.)	short untimed pause
(1.5)	time pause in seconds and tenths of seconds
((head nod))	double parentheses for description of events
[((gesture))	left bracket double parentheses for embodied action
word	italicizing for stress
lo:::ng	colon(s) for vowel lengthening
bold	louder than surrounding talk
bold italics	loudness+stress
=	equal sign for latched utterances with no pause
look-	dash for cut off utterances
[((extends open palm))	below brackets for overlapping speech/gesture
(B)	(B) in parentheses for beat underneath its accompanying word or phrase

References

Aikhenvald, Alexandra. 2004. *Evidentiality.* New York: Oxford U. Press.
Alexander, Jeffery. 2012. *Trauma: A Social Theory.* Cambridge: Polity Press.
Alibali, Martha & Alyssa DiRusso. 1999. The function of gesture in learning to count: More than keeping track. *Cognitive Development* 14. 37–56.
Atkinson, John. 1984. *Our Master's Voices.* London: Routledge.
Atkinson, John & Paul Drew. 1979. *Order in Court.* New York: MacMillian.
Bauman, Richard & Charles Briggs. 2003. *Voices of Modernity: Language Ideologies and the Politics of Inequality.* Cambridge: Cambridge University Press.
Besnier, Niko. 1990. Language and affect. *Annual Review of Anthropology* 19. 419–451.
Biber, Douglas, Stig Johansson, Geoffrey Leech, Susan Conrad & Edward Finegan. 1999. *Grammar of Spoken and Written English.* New York: Longman.
Bietti, Lucas. 2010. Sharing memories, family conversation and interaction. *Discourse and Society* 21 (5). 499–523.
Bulwer, John. 2003 [1644]). *Chirologia or the Natural Language of the Hand.* Whitefish, MT: Kessinger Publishing.
Burgess, Ann & Linda Holmstrum. 1971. Rape trauma syndrome. *American Journal of Psychiatry* 131 (9). 981–986.
Calbris, Genevieve. 2011. *Elements of Meaning in Gesture.* Amsterdam: John Benjamins.
Chaemsaithong, Krisda. 2017a. Evaluative stancetaking in courtroom opening statements. *Folia Linguistica* 5 (1). 103–132.
Chaemsaithong, Krisda. 2017b. Speech reporting in courtroom opening statements. *Journal of Pragmatics* 119. 1–14.
Chaemsaithong , Krisda. 2018. Use of voices in legal opening statements. *Social Semiotics* 28 (1). 90–107.
Clarke, Thurston. 2008. *The Last Campaign: Robert F. Kennedy and 82 Days that Inspired America.* New York: Holt.
Clark, Herbert & Richard Gerrig.1990. Quotations as demonstrations. *Language* 66. 764–805.
Clayman, Steve & John Heritage. 2002. *The News Interview.* Cambridge: Cambridge U. Press.
Cooperrider, Kensy & Rafael Nunez. 2009. Across time, across the body: Traversal temporal gestures. *Gesture* 9 (2). 181–206.
Cooperrider, Kensy, Natasha Abner & Susan Goldin-Meadow. 2018. The palm-up puzzle: Meanings and Origins of a widespread form in gesture and sign. *Frontiers in Communication* 1–16.
Corning Amy & Howard Schuman. 2015. *Generations and Collective Memory.* Chicago: University of Chicago Press.
Cossins, Anne. 2020. *Closing the Justice Gap for Adult and Child Sexual Assault.* London: Palgrave Macmillan.
Cotterill, Janet. 2003. *Language and Power in Court: A Linguistic Analysis of the O.J. Simpson Trial.* New York: Palgrave.
Craig, Elaine. 2021. *Putting Trials on Trial: Sexual Assault and the Failure of the Legal Profession.* Montreal: McGill-Queen's University Press.
de Jorio, Andrea. 2000. *Gesture in Naples and Gesture in Classical Antiquity.* A translation of La Mimica Degli Antichi Investigata Nel Gestire Napoletano (1832). Bloomington: Indiana University Press.
Duranti, Alessandro. 1997. *Linguistic Anthropology.* Cambridge: Cambridge University Press.
Ehrlich, Susan 2001. *Representing Rape: Language and Sexual Consent.* London: Routledge.

Ehrlich, Susan & Alice Freed. 2010. The function of questions in institutional discourse. In Susan Erhlich & Alice Freed (eds.), *Why Do You Ask: The Function of Questions in Institutional Discourse*, 3–29. New York: Oxford U. Press.
Ekman, Paul & Wallace Friesen. 2003. *Unmasking the Face: A Guide to Recognizing Emotions from Facial Clues.* Cambridge, MA: Major Books.
Ekman, Paul. 2003. *Emotions Revealed.* New York: St. Martins Press.
Erickson, Frederick & Jeffery Schultz. 1982. *Counselors as Gatekeeper: Social Interaction in Interviews.* New York: Academic Press.
Eyerman Roy. 2011 *The Cultural Sociology of Political Assassination.* New York: Palgrave.
Eyerman Roy. 2012. Cultural trauma: Emotion and narration. In Jeffrey Alexander, Ronald Jacobs & Philip Smith (eds.), *The Oxford Handbook of Cultural Sociology*, 564–582. New York: Oxford University Press.
Ferré, Gaelle. 2011. Functions of three open-palm gestures. *Multimodal Communication* 1 (1). 5–20.
Fleming, Luke & Michael Lempert. 2014. Poetics and performativity. In Nick Enfield, Paul Kockelman & Jack Sidnell (eds.), *The Cambridge Handbook of Linguistic Linguistic Anthropology*, 485–515. Cambridge: Cambridge University Press.
French, Brigittine. 2012. The semiotics of collective memories. *Annual Review of Anthropology* 41. 337–353.
Friedman, Lawrence. 1977. *Law and Society: An Introduction.* Englewood Cliffs, NJ.: Prentice-Hall.
Gal, Susan & Judith Irvine. 2019. *Signs of Difference.* Cambridge: Cambridge University Press.
Gibbons, John. 2003. *Forensic Linguistics.* Oxford: Blackwell.
Giddens, Anthony. 1979. *Central Problems in Social Theory.* Berkeley: University of California Press.
Giddens, Anthony. 1984. *The Constitution of Society.* Berkeley: University of California Press.
Gilbert, Kristin & Gregory Matoesian. 2015. Multimodal action and speaker positioning in closing argument. *Multimodal Communication* 4(2). 93–112.
Gilbert, Kristin & Gregory Matoesian. 2021. *Multimodal Performance and Interaction in Focus Groups.* Amsterdam: John Benjamins.
Ginns, Paul, Hu, Fang-Tzu, Erin Byrne & Janette Bobis. 2015. Learning by tracing worked examples. *Applied Cognitive Psychology* 30. 160–169.
Givon, Thomas. 1978. Negation in language: Pragmatics, function, ontology. In Peter Cole (ed.). *Syntax and Semantics* (vol. 9), 69–112. New York: Academic Press.
Goffman, Erving. 1981 *Forms of Talk.* Philadelphia: University of Pennsylvania Press.
Goldin-Meadow, Susan. 2003. *Hearing Gesture.* Cambridge MA: Harvard University Press.
Goldin-Meadow, Susan. *in press.* Nonverbal communication: The hand's role in talking and thinking. In Richard Lerner (ed.), *Handbook of Child Psychology and Developmental Science* (vol 2). New York: Wiley.
Goodwin, Charles. 1994. Professional vision. *American Anthropologist* 95(3). 606–633.
Goodwin, Charles. 2003. Pointing as situated practice. In Kita Sotaro (ed.), *Pointing: Where Language, Culture, and Cognition Meet*, 217–241. Mahwah: Lawrence Erlbaum.
Halbwachs, Maurice. 1992. *On Collective Memory.* Edited, Translated, and with an Introduction by Lewis Coser. Chicago: University of Chicago Press.
Harrison, Simon. 2018. *The Impulse to Gesture: Where Language, Minds, and Bodies Intersect.* Cambridge: Cambridge University Press.
Haydock, Roger & John Sonsteng. 1990. *Trial Theories, Tactics and Techniques.* St. Paul, Minnesota: West.
Heath, Christian, John Hindmarsh & Paul Luff. 2010. *Video in Qualitative Research.* London: Sage.
Heffer, Chris. 2005. *The Language of the Jury Trial. A Corpus-Aided Linguistic Analysis of Legal-Lay Discourse.* New York: Palgrave.

Heffer, Chris. 2010. Narrative in the trial: Constructing crime stories in court. In Malcolm Coulthard & Allison Johnson (eds.), *Routledge Handbook of Forensic Linguistics*, 199–217. New York: Routledge.

Heffer, Chris, Francis Rock & John Conley (eds.). 2013. *Legal-Lay Communication*. New York: Oxford University Press.

Hepburn, Alexa & Jonathan Potter. 2012. Crying and crying responses. In Anssi Perakyla & Marja-Leena Sorjonen (eds.), *Emotion in Interaction*, 195–211. New York: Oxford University Press.

Heritage, John. 1984. *Garfinkel and Ethnomethodology*. Cambridge: Polity Press.

Heritage, John. 2010. Questions in medicine. In Susan Erhlich & Alice Freed (eds.), *Why Do You Ask: The Function of Questions in Institutional Discourse*. 42–68. New York: Oxford U. Press.

Heritage, John. & David Greatbatch. 1986. Generating applause. A study of rhetoric and response at party political conferences. *American Journal of Sociology* 92 (1). 110–157.

Heritage, John & Steve Clayman. 2010. *Talk in Action: Interactions, Identities, and Institutions.* Chicester, UK: Wiley Blackwell.

Hibbitts, Bernard. 1995. Making motions: The embodiment of law in gestures. *Journal of Contemporary Legal Issues* 6. 51–81.

Hobbs, Pamela. 2008. Discourse in the law. In Wolfgang Donsbach (ed.), *The International Encyclopedia of Communication,* (vol. 4), 239–240. Oxford: Wiley-Blackwell.

Holt, Elizabeth & Alison Johnson. 2010. Legal Talk. In Malcolm Coulthard & Allison Johnson (eds.), *Routledge Handbook of Forensic Linguistics,* 21–36. New York: Routledge.

Horn, Laurence. 1979. *A Natural History of Negation.* Chicago: U. of Chicago Press.

Hu, Fang-Tzu, Paul Ginns & Janette Bobis. 2014. Does tracing worked examples enhance geometry learning. *Australian Journal of Educational and Developmental Psychology* 14. 45–49.

Hu, Fang-Tzu, Paul Ginns & Janette Bobis. 2015. Getting the point: Tracing worked examples enhances learning. *Learning and Instruction* 35. 85–93.

Huber, Judith. 2017. *Motion and the English Verb.* New York: Oxford University Press.

Jaffe, Alexandra. 2009. Introduction: The sociolinguistics of stance. In Alexandra Jaffe (ed.), *Stance: Sociolinguistic Perspectives,* 3–28. New York: Oxford U. Press.

Jackendoff, Ray. 2002. *Foundations of Language.* New York: Oxford University Press.

Jakobson, Roman. 1960. Closing Statement: Linguistics and Poetics. In Thomas Sebeok (ed.), *Style in Language*, 398–429. Cambridge, Massachusetts: MIT Press.

Klippel, Alexander, Thora Tenbrink & Daniel Montello. 2013. The role of structure and function in the conceptualization of direction. In Mila Vulchanova & Emile van der Zee (eds.), *Motion Encoding in Language and Space,* 102–110. New York: Oxford University Press.

Kendon, Adam. 2004. *Gesture: Visible Action as Utterance.* Cambridge: Cambridge University Press.

Komter, Martha. 1997. *Dilemmas in the Courtroom.* New York: L. Erlbaum.

Komter, Martha. 2000. The power of legal language: The significance of small activities for large problems. *Semiotica* 131-3/4. 415–428.

Komter, Martha. 2019. *The Suspect's Statement.* Cambridge: Cambridge University Press.

Krahmer, Emiel & Marc Swerts. 2007. The effects of visual beats on prosodic prominence: Accoustic analyses, auditory perception and visual perception. *Journal of Memory and Language* 57. 396–414.

Lakoff, George & Mark Johnson. 1980. *Metaphors We Live By.* Chicago: University of Chicago Press.

Landsman, Stephan. 1984. *The Adversary System.* Ann Arbor, Mi: AEI Studies.

Lees, Susan. 2002. *Carnal Knowledge: Rape on Trial.* London: Penquin.

Lempert, Michael. 2011. Barack Obama, being sharp: Indexical order in the pragmatics of precision-grip gesture. *Gesture* 11. 241–270.

Lempert, Michael. 2012. Indirectness. In Christina Paulston, Scott Kiesling & Elizabeth Rangel (eds.), *The Handbook of Intercultural Communication,* 180–204. Oxford: Wiley-Blackwell.

Lempert, Michael. 2017. Uncommon resemblance: Pragmatic affinity in political gesture. *Gesture* 16. 35–67.
Lempert, Michael. 2018. On the pragmatic poetry of pose: Gesture, parallelism, politics. *Signs and Society* 6 (1). 120–146.
Lempert, Michael & Michael Silverstein. 2012. *Creatures of Politics.* Bloomington: Indiana University Press.
Levin, Beth. 1993. *English Verb Classes and Alternations.* Chicago: University of Chicago Press.
Maricchiolo, Fridanna, Augusto Gnisci, Marino Bonaiuto & Gianluca Ficca. 2009. Effects of different types of hand gestures in persuasive speech on receivers' evaluations. *Language and Cognitive Processes* 24. 239–266.
Matoesian, Gregory. 1993. *Reproducing Rape: Domination through Talk in the Courtroom.* Chicago: University of Chicago Press.
Matoesian, Gregory. 1995. Language, law, and society: Policy implications of the Kennedy-Smith rape trial. *Law and Society Review* 29 (4). 669–702.
Matoesian, Gregory. 2001. *Law and the Language of Identity.* Oxford: Oxford University Press.
Matoesian, Gregory. 2005. Struck by speech revisited: Embodied stance in jurisdictional discourse. *Journal of Sociolinguistics* 9. 167–194.
Matoesian, Gregory M. 2010. Multimodal aspects of victim narration in direct examination. In Malcolm Coulthard & Allison Johnson (eds.), *Routledge handbook of forensic linguistics*, 541–557. New York: Routledge.
Matoesian, Gregory. & Kristin Gilbert. 2016. Multifunctionality of beat gestures and material conduct in closing argument. *Gesture* 15. 79–114.
Matoesian, Gregory. & Kristin Gilbert. 2018. *Multimodal Conduct in the Law:* Cambridge: Cambridge University Press.
Matoesian, Gregory. & Lynn Taylor. 1983. Academic exploitation of the sexually abused: The third time a woman is raped. Paper presented at the American Society of Criminology Meetings, Denver.
Mauet, Thomas. 1996. *Trial Techniques* (4th edition). Boston, MA: Little, Brown and Co.
Mauet, Thomas. 2017. *Trial Techniques* (10th edition). New York: Aspen.
Mauet, Thomas & Stephen Easton. 2021. *Trial Techniques and Trials.* (11th edition). New York: Wolters Kluwer.
McHoul, Alexander. 1978. The organization of turns at formal talk in the classroom. *Language in Society* 7. 183–213.
McNeill, David. 1992. *Hand and Mind: What Gestures Reveal About Thought.* Chicago: University of Chicago Press.
McNeill, David. 2005. *Gesture and Thought.* Chicago: University of Chicago Press.
McNeill, David. 2006. Gesture and Communication. In Keith Brown (ed.), *Encyclopedia of Linguistics* (2nd edition), 58–67. New York: Elsevier.
McNeill, David 2012. *How Language Began: Gesture and Speech in Human Evolution.* New York: Cambridge University Press.
Mertz, Elizabeth. 2007. *The Language of Law School.* New York: Oxford University Press.
Morris, Desmond. 1977. *Manwatching: A Field Guide to Human Behaviour.* London: Jonathan Cape.
Müller, Cornelia. 2004. Forms and uses of the palm up open hand: A case of a gesture family? In Cornela. Muller and Roland Posner (eds.), *The Semantics and Pragmatics of Everyday Gestures*, 233–256. Berlin: Weidler Buchverlag.
Müller, Cornelia. 2008. What gestures reveal about the nature of metaphor. In Alan Cienki & Cornelia Müller (eds.), *Metaphor and Gesture*, 219–245. Philadelphia: John Benjamins.

Müller, Cornelia. 2014. Ring-gestures across cultures and times. In Cornelia Müller, Alan Cienki, Ellen Fricke, Silva Ladewig, David McNeill,& Jana Bressem (eds.), *Body – Language – Communication* (vol. 2), 1511–1522. Berlin: DeGruyter Mouton.

Nordlund, Taru & Heli Pekkarinen. 2014. Grammaticalisation of Finnish stance adverbial *muka*,'as if, supposedly, allegedly'. In Irma Taavitsainen, Andreas Jucker & Jukka Tuominen (eds.), *Diachronic Corpus Pragmatics*, 53–76. Amsterdam: John Benjamins.

Ochs, Elinor & Bambi Schieffelin. 1989. Language has a heart. *Text* 9. 7–25.

Perrin, Larry, Harry Caldwell & Carol Chase. 2003. *The Art and Science of Trial Advocacy*. Cincinnati, OH: Anderson.

Philips, Susan. 1986. Reported speech as evidence in an American trial. In Deborah Tannen & Jane Alatis (eds.), *Language and Linguistics: The Interdependence of Theory, Data and Application*, 154–179. Washington DC: Georgetown University Press.

Quintilian, Marcus. 2001. *The Orator's Education, trans. D.A. Russell*. Cambridge, MA: Harvard University Press.

Raymond, Geoffrey. 2006. Questions at Work: Yes/No Type Interrogatives in Institutional Contexts. In Paul Drew, Geoffrey Raymond & Darin Weinberg (eds.), *Talk and Interaction in Social Research Methods*, 115–134. London: Sage.

Roseano, Paolo, Montserrat Gonzalez, Comes Borras & Pilar Prieto. 2016. Communicating epistemic stance: How speech and gesture patterns reflect epistemicity and evidentiality. *Discourse Process* 53 (3). 135–174.

Rosulek, Laura. 2008 Manipulative silence and social representation in the closing arguments of a child sexual abuse case. *Text & Talk* 28. 529–550.

Rosulek, Laura. 2010. Prosecution and defense closing speeches: The creation of contrastive closing arguments. In Malcolm Coulthard, & Alison Johnson (eds.), *The Routledge Handbook of Forensic Linguistics*, 218–230. New York, New York: Routledge.

Sacks, Harvey. 1988. *Lectures in Conversation*. New York: Blackwell.

Sauer, Beverly. 2003. *The Rhetoric of Risk: Technical Documentation in Hazardous Environments*. Malwah, NJ: Lawrence Erlbaum.

Schwartz Barry. 1982. The social context of commemoration: A study in collective memory. *Social Forces* 61 (2). 374–402.

Schwartz, Barry. 2007. Collective memory. In George Ritzer (ed.), *The Blackwell Encyclopedia of Sociology*. 588–590. Oxford: Blackwell.

Sidnell, Jack. 2006. Coordinating gesture, talk, and gaze in reenactments. *Research on Language and Social Interaction* 39. 377–409.

Silverstein, Michael. 1976. Language structure and linguistic ideology. In Paul Clyne, William Hanks & Carol Hofbauer (eds.), *The Elements: A Parasession on Linguistic Units and Levels*. 193–247. Chicago: University of Chicago Press.

Silverstein, Michael. 1981. The limits of awareness. *Sociolinguistic Working Paper* Number 84. Southwest Educational Development Lab., Austin, TX.

Silverstein, Michael. 1985. On the pragmatic "poetry" of prose: Parallelism, repetition, and cohesive structure in the time course of dyadic conversation. In Deborah Schiffrin (ed.), *Meaning, Form, and Use in Context: Linguistic Applications*, 181–198. Washington DC: Georgetown University Press.

Silverstein, Michael. 1998. The improvisational performance of culture in realtime discursive practice. In Keith Sawyer (ed.), *Creativity in Performance*, 265–312. Greenwich, CT: Ablex.

Silverstein, Michael. 1993. Metapragmatic discourse and the metapragmatic function. In John Lucy (ed.), *Reflexive Language*, 33–58. New York: Cambridge University Press. 33–58.

Silverstein, Michael. 2004. Cultural concepts and the language-culture nexus. *Current Anthropology* 45 (5). 621–652.

Silverstein, Michael. 2014. Denotation and the pragmatics of language. In Nick Enfield, Paul Kockelman & Jack Sidnell (eds.), *Handbook of Linguistic Anthropology*, 128–157. Cambridge: Cambridge University Press.

Simpson, Alfred. 1988. *Invitation to Law*. Oxford: Blackwell.

Smelser, Neil. 2004. Psychological trauma and cultural trauma. In Jeffery Alexander, Ron Eyerman, Bernard Giesen, Neil Smelser & Piotr Sztompka (eds.), *Cultural Trauma and Collective Identity*, 31–59. Berkeley: University of California Press.

Smith, Olivia. 2018. *Rape Trials in England and Wales: Observing Justice and Rethinking Rape Myths*. New York: Palgrave-Macmillan.

Stasch, Rubert. 2011. Ritual and oratory revisited: The semiotics of effective action. *Annual Review of Anthropology* 40. 159–74.

Streeck, Jürgen. 1993. Gesture as communication I: Its coordination with gaze and speech. *Communication Monographs* 60. 275–299.

Streeck, Jürgen. 2008. Gesture in political communication: A case study of the democratic presidential candidates during the 2004 primary campaign. *Research on Language and Social Interaction* 41 (2).154–186.

Streeck, Jürgen. 2009. *Gesturecraft: The Manufacture of Meaning*. Amsterdam: John Benjamins.

Stygall, Gail. 2012. Discourse in the U.S. courtroom. In Lawrence Solan & Peter Tiersma (eds.), *Oxford Handbook of Language and Law*. 369–380. New York: Oxford University Press.

Talmy, Leonard. 1985. Lexicalization patterns: semantic structure in lexical forms. In Timothy Shopen (ed.), *Language Typology and Syntactic Description. Volume III: Grammatical Categories and the Lexicon*, 57–149. Cambridge: Cambridge University Press.

Tanford, John. 1983. *The Trial Process: Law, Tactics and Ethics*. Charlottesville, Virginia: Michie.

Tannen, Deborah .1989. *Talking Voices*. New York: Cambridge University Press.

Taslitz, Andrew. 1999. *Rape and the Culture of the Courtroom*. New York: New York University Press.

Tavarez, David. 2014. Ritual language. In Nick Enfield, Paul Kockelman & Jack Sidnell (eds.), *The Cambridge Handbook of Linguistic Anthropology*, 516–536. Cambridge: Cambridge University Press.

Taylor, John. 1993. Prepositions: Patterns of polysemization and strategies of disambiguation. In Cornelia Zelinsky-Wibbelt (ed.), *The Semantics of Prepositions: From Mental Processing to Natural Language Processing*, 151–175. Berlin: Mouton de Gruyter.

Tebendorf, Sedinha. 2014. Pragmatic and Metaphoric – Combining Functional with Cognitive Approaches in the Analysis of the "Brushing Aside Gesture". In Cornelia Müller, Alan Cienki, Ellen Fricke, Silva Ladewig, David McNeill & Jana Bressem (eds.), *Body – Language – Communication* (vol.2), 1540–1558. Berlin: De Gruyter Mouton.

Temkin, Jennifer, Jacqueline Gray & Jastine Barrett. 2018. Different functions of rape myth use in court: Findings from a trial observation study. *Feminist Criminology* 13 (2). 205–226.

Tiersma, Peter. 1999. *Legal Language*. Chicago: University of Chicago Press.

Tracy, Karen & Jessica Robles. 2009. Questions, questioning, and institutional practices. *Discourse Studies* 11. 131–152.

Trinch, Shonna. 2003. *Latinas' Narratives of Domestic Abuse*. Amsterdam: John Benjamins.

Wermeskerken, Margot, Nathalie Fijan, Charly Eielts & Wim Pouw. 2016. Observation of depictive versus tracing gestures selectively aids verbal versus visual-spatial learning in primary school children. *Applied Cognitive Psychology* 30. 806–814.

Wilce, James. 2009. *Language and Emotion*. Cambridge University Press.

Wilce, James. 2017. *Culture and Communication*. Cambridge: Cambridge University Press.

Wodak, Ruth & Rudolf de Cillia. 2007 Commemorating the past: The discursive construction of official narratives about the 'Rebirth of the Second Austrian Republic' *Discourse and Communication* 1 (3). 337–363.

Woolard, Kathryn. 2021. Language ideology. In James Stanlaw (ed.), *The International Encyclopedia of Linguistic Anthropology*, 1–21. New York: John Wiley and Sons.

Wortham, Stanton. 2001. *Narratives in Action.* New York: Teachers College Press.

Index

Acoustic gesture 60, 135
Affect 25, 70
Audio-visual data 14

Beats 8, 23, 33–34, 38–40, 48, 53, 56, 66, 71, 72, 74–76, 78, 91, 108, 115, 117–118, 135
Bodily quotation 114

Closing argument 90
Collective memory 64, 65, 70, 76, 81
Counting 109
Court reporter 35
Cross-modal 78
Cultural trauma 70, 73

Data 13
Demonstrations/enactments 113, 118, 121, 123, 125–128
Denotational text 36
Diagram 46–47, 53
Dialect of the fingers 1, 38, 90, 109, 132, 134
Direct quotes 34, 42, 112–113, 140
Durational verb 53, 54

Embodied resistance ideology 97, 110
Emotion 83, 86–87
Epistemics 55
Evidence expert 20–21
Evidential 55
Exhibits 43

Facial expressions 10
Footing 126
Forensic linguistics 1, 6

Gaze 10, 85, 95–96, 101, 127
Gestural aspect 126
Gesture 7, 8, 33, 34, 66, 78, 83, 91, 105, 122
Gesture hold 8, 99, 107
Gesture stroke 8, 99, 107

Head tilt 25, 37

Inconsistency 129–130
Interactional text 36
Intradigital beats 38–40, 102
Interdigital beats 38–40, 92–100, 103, 109
Interdiscursivity 112–113, 116
Involvement 26–28

Kennedy Smith trial 1

Legal identity 20
Legal ritual 13
Linguistic ideology 129–130

Material conduct 10, 44, 47, 49
Metaphor 83, 91
Methodology 13–15
Micro-ethnography 14
Motion 10, 44, 52, 54, 59
Multimodal conduct 6, 11, 44, 96

Neutral translator 136

Objections 4, 5, 15, 18–20, 35, 134
Objection conference 32
Opening statement 42, 63
Oratory 1, 17, 132

Participation 19, 23, 24, 29, 36, 40, 116, 133
Path preposition 59
Poetics 11–13, 39, 51, 52, 62, 73, 74, 76, 118, 123, 132, 135
Power 109–110
Pre-recorded questions 136
Precision ring gesture 33

Questioning 4, 5, 18, 29, 133, 138

Rape myths 141
Rape reform 135–141
Residual semanticity 8, 108, 135
Ritual oratory 73, 79–80

Search verb 62
Sexual history 141
Speaker centric ideology 20
Speech exchange system 2, 6, 41
Speech-gesture ensemble 8, 72, 97
Stance 23–24, 33, 47, 133

Tracing 56–63, 134
Transcript conventions 143

Utmost resistance ideology 97

www.ingramcontent.com/pod-product-compliance
Lightning Source LLC
Chambersburg PA
CBHW050527170426
43201CB00013B/2115